Lecture Notes in Social Networks

Lecture Notes in Social Networks (LNSN) comprises volumes covering the theory, foundations and applications of the new emerging multidisciplinary field of social networks analysis and mining. LNSN publishes peer-reviewed works (including monographs, edited works) in the analytical, technical as well as the organizational side of social computing, social networks, network sciences, graph theory, sociology, Semantics Web, Web applications and analytics, information networks, theoretical physics, modeling, security, crisis and risk management, and other related disciplines. The volumes are guest-edited by experts in a specific domain. This series is indexed by DBLP. Springer and the Series Editors welcome book ideas from authors. Potential authors who wish to submit a book proposal should contact Christoph Baumann, Publishing Editor, Springer e-mail: Christoph.Baumann@springer.com

More information about this series at http://www.springer.com/series/8768

Mohammad A. Tayebi • Uwe Glässer
David B. Skillicorn

Editors

Open Source Intelligence and Cyber Crime

Social Media Analytics

 Springer

Editors
Mohammad A. Tayebi
School of Computing Science
Simon Fraser University
Burnaby, BC, Canada

Uwe Glässer
School of Computing Science
Simon Fraser University
Burnaby, BC, Canada

David B. Skillicorn
School of Computing
Queen's University
Kingston, ON, Canada

ISSN 2190-5428   ISSN 2190-5436   (electronic)
Lecture Notes in Social Networks
ISBN 978-3-030-41253-1   ISBN 978-3-030-41251-7   (eBook)
https://doi.org/10.1007/978-3-030-41251-7

This Springer imprint is published by the registered company Springer Nature Switzerland AG
The registered company address is: Gewerbestrasse 11, 6330 Cham, Switzerland

Contents

Protecting the Web from Misinformation

Francesca Spezzano and Indhumathi Gurunathan

Abstract Nowadays, a huge part of the information present on the Web is delivered through Social Media and User-Generated Content (UGC) platforms, such as Quora, Wikipedia, YouTube, Yelp, Slashdot.org, Stack Overflow, Amazon product reviews, and much more. Here, many users create, manipulate, and consume content every day. Thanks to the mechanism by which anyone can edit these platforms, its content grows and is kept constantly updated. However, malicious users can take advantage of this open editing mechanism to introduce misinformation on the Web.

In this chapter, we focus on Wikipedia, one of the main UCC platform and source of information for many, and study the problem of protecting Wikipedia articles from misinformation such as vandalism, libel, spam, etc. We address the problem from two perspectives: detecting malicious users to block such as spammers or vandals and detecting articles to protect, i.e., placing restrictions on the type of users that can edit an article. Our solution does not look at the content of the edits but leverages the users' editing behavior so that it generally results applicable to many languages. Our experimental results show that we are able to classify (1) article pages to protect with an accuracy greater than 92% across multiple languages and (2) spammers from benign users with 80.8% of accuracy and 0.88 mean average precision.

The chapter also defines different types of misinformation that exist on the Web and provides a survey of the methods proposed in the literature to prevent misinformation on Wikipedia and other platforms.

1 Introduction

Nowadays, a huge part of the information present on the Web is delivered through Social Media such as Twitter, Facebook, Instagram, etc., and User-Generated Content (UGC) platforms, such as Quora, Wikipedia, YouTube, Yelp, Slashdot.org,

F. Spezzano (✉) · I. Gurunathan
Computer Science Department, Boise State University, Boise, ID, USA
e-mail: francescaspezzano@boisestate.edu; indhumathigurunathan@u.boisestate.edu

© Springer Nature Switzerland AG 2020
M. A. Tayebi et al. (eds.), *Open Source Intelligence and Cyber Crime*, Lecture Notes in Social Networks, https://doi.org/10.1007/978-3-030-41251-7_1

1

Stack Overflow, Amazon product reviews, and many others. Here, users create, manipulate, and consume content every day. Thanks to the mechanism by which anyone can edit these platforms, its content grows and is kept constantly updated.

Unfortunately, Web features that allow for such openness have also made it increasingly easy to abuse this trust, and as people are generally awash in information, they can sometimes have difficulty discerning fake stories or images from truthful information. They may also lean too heavily on information providers or social media platforms such as Facebook to mediate even though such providers do not commonly validate sources. For example, most high school teens using Facebook do not validate news on this platform. The Web is open to anyone, and malicious users shielded by their anonymity threaten the safety, trustworthiness, and usefulness of the Web; numerous malicious actors potentially put other users at risk as they intentionally attempt to distort information, manipulate opinions and public response. Even worse, people can get paid to create fake news and spam reviews, influential bots can easily create it, and misinformation spreads so fast that is too hard to control. Impacts are already destabilizing the U.S. electoral system and affecting civil discourse, perception, and actions since what people read on the Web and events they think happened may be incorrect, and people may feel uncertain about their ability to trust it.

Misinformation can manifest in multiple forms such as vandalism, spam, rumors, hoaxes, fake news, clickbaits, fake product reviews, etc. In this chapter, we start by defining misinformation and describing different forms of misinformation that exist nowadays on the Web. Next, we focus on how to protect the Web from misinformation and provide a survey of the methods proposed in the literature to detect misinformation on social media and user-generated contributed platforms. Finally, we focus on Wikipedia, one of the main UGC platform and source of information for many, and study the problem of protecting Wikipedia articles from misinformation such as vandalism, libel, spam, etc. We address the problem from two perspectives: detecting malicious users to block such as spammers or vandals and detecting articles to protect, i.e., placing restrictions on the type of users that can edit an article. Our solution does not look at the content of the edits but leverages the users' editing behavior so that it generally results applicable to many languages. Our experimental results show that we are able to classify (1) article pages to protect with an accuracy greater than 92% across multiple languages and (2) spammers from benign users with 80.8% of accuracy and 0.88 mean average precision. Moreover, we discuss one of the main side effects of deploying anti-vandalism tools on Wikipedia, i.e. a low rate of newcomers retention, and an algorithm we proposed to early detect whether or not a user will become inactive and leave the community so that recovery actions can be performed on time to try to keep them contributing longer.

This chapter differs from the one by Wu et al. [1] because we focus more on the Wikipedia case study and how to protect this platform from misinformation, while Wu et al. mainly deal with rumors and fake news identification and intervention. Other related surveys are the one by Shu et al. [2] that focuses specifically on fake news, the work by Zubiaga [3] that deals with rumors, and the survey by Kumar and Shah [4] on fake news, fraudulent reviews, and hoaxes.

2 Misinformation on the Web

According to the Oxford dictionary, *misinformation* is "false or inaccurate information, especially that which is deliberately intended to deceive". These days, the massive growth of the Web and social media has provided fertile ground to consume and quickly spread the misinformation without fact-checking. Misinformation can assume many different forms such as *vandalism, spam, rumors, hoaxes, counterfeit websites, fake product reviews, fake news, etc.*

Social media and user-generated content platforms like Wikipedia and Q&A websites are more likely affected by vandalism, spam, and abuse of the content. **Vandalism** is the action involving deliberate damage to others property, and Wikipedia defines vandalism on its platform as "the act of editing the project in a malicious manner that is intentionally disruptive" [5]. Beyond Wikipedia, other user-generated content platforms on the Internet got affected by vandalism. For example, editing/down-voting other users content in Q&A websites like Quora, Stack Overflow, Slashdot.org, etc. Vandalism can also happen on social media such as Facebook. For instance, the Martin Luther King, Jr.'s fan page was vandalized in Jan 2011 with racist images and messages.

Spam is, instead, a forced message or irrelevant content sent to a user who would not choose to receive it. For example, sending email to a bulk of users, flooding the websites with commercial ads, adding external link to the articles for promoting purposes, improper citations/references, spreading links created with the intent to harm, mislead or damage a user or stealing personal information, likejacking (tricking users to post a Facebook status update for a certain site without the user's prior knowledge or intent), etc.

Wikipedia, like most forms of online social media, receives continuous spamming attempts every day. Since the majority of the pages are open for editing by any user, it inevitably happens that malicious users have the opportunity to post spam messages into any open page. These messages remain on the page until they are discovered and removed by another user. Specifically, Wikipedia recognizes three main types of spam, namely "advertisements masquerading as articles, external link spamming, and adding references with the aim of promoting the author or the work being referenced" [6].

User-generated content platforms define policies and methods to report vandalism, and spam and the moderation team took necessary steps like warning the user, blocking the user from editing, collapse the content if it is misinformation, block the question from visible to other users, or ban the user from writing/editing answers, etc. These sites are organized and maintained by the users and built as a community. So the users have responsibilities to avoid vandalism and make it as a knowledgeable resource to others. For example, the Wikipedia community adopts several mechanisms to prevent damage or disruption to the encyclopedia by malicious users and ensure content quality. These include administrators to ban or block users or IP addresses from editing any Wikipedia page either for a finite amount of time or indefinitely, protecting pages from editing, or detecting damaging content to be reverted through dedicated bots [7, 8], monitoring recent changes, or having watch-lists.

Slashdot gives moderator access to its users to do jury duty by reading comments and flag it with appropriate tags like Offtopic, Flamebait, Troll, Redundant, etc. Slashdot editors also act as moderators to downvote abusive comments. In addition to that, there is an "Anti" symbol present for each comment to report spam, racist ranting comments, etc. Malicious users can also act protected by anonymity. In Quora, if an anonymous user vandalizes the content, then a warning message is sent to that user's inbox without revealing the identity. If the particular anonymous user keeps on abusing the content, then Quora moderator revokes the anonymity privileges of that user.

Online reviews are not free from misinformation either. For instance, on Amazon or Yelp, it is frequent to have spam paid reviewers writing **fraudulent reviews** (or opinion spam) to promote or demote products or businesses. Online reviews help customers to make decisions on buying the products or services, but when the reviews are manipulated, it will impact both customers and business [9]. Fraudulent reviewers post either positive review to promote the business and receive something as compensation, or they write negative reviews and get paid by the competitors to create damage to the business. There are some online tools like `fakespot.com` and `reviewmeta.com` that analyze the reviews and helps to make decisions. But in general, consumers have to use some common sense to not fall for fraudulent reviews and do some analysis to differentiate the fake and real reviews. Simple steps like verifying the profile picture of the reviewer, how many other reviews they wrote, paying attention to the details, checking the timestamp, etc., will help to identify fraudulent reviews.

Companies also take some actions against fraudulent reviews. Amazon sued over 1000 people who posted fraudulent reviews for cash. It is also suspending the sellers and shut-downing their accounts if they buy fraudulent reviews for their products. They rank the reviews and access the buyer database to mark the review as "Verified Purchase" meaning that the customer who wrote the review also purchased the item at Amazon.com. Yelp has an automated filtering software that is continuously running to examine each review recommend only useful and reliable reviews to its consumers. Yelp also leverages the crowd (consumer community) to flag suspicious reviews and takes legal action against the users who are buying or selling reviews.

Fake news is low-quality news that is created to spread misinformation and misleading readers. The consumption of news from social media is highly increased nowadays so as spreading of fake news. According to the Pew research center [10], 64% of Americans believe that fake news causes confusion about the basic facts of current events. A recent study conducted on Twitter [11] revealed that fake news spread significantly more than real ones, in a deeper and faster manner and that the users responsible for their spread had, on average, significantly fewer followers, followed significantly fewer people, were significantly less active on Twitter. Moreover, bots are equally responsible for spreading real and fake news, and then the considerable spread of fake news on Twitter is caused by human activity.

Fact-checking the news is important before spreading it on the Web. There are a number of news verifying websites that can help consumers to identify fake news by making more warranted conclusions in a fraction of the time. Some examples of fact-checkers are `FactCheck.org`, `PolitiFact.com`, `snopes.com`, or `mediabiasfactcheck.com`.

Beyond fact-checking, consumers should also be responsible for [12]:

1. *Read more than the headline*—Often fake news headlines are sensational to provoke readers emotions that help the spread of fake news when readers share or post without reading the full story.
2. *Check the author*—The author page of the news website provides details about the authors who wrote the news articles. The credibility of the author helps to measure the credibility of the news.
3. *Consider the source*—Before sharing the news on social media, one has to ensure the source of the articles, verify the quotes that the author used in the article. Also, a fake news site often has strange URL's.
4. *Check the date*—Fake news sometimes provides links to previously happened incidents to the current events. So, one needs to check the date of the claim.
5. *Check the bias*—If the reader has opinion or beliefs to one party, then they tend to believe biased articles. According to a study done by Allcott and Gentzkow [13], the right-biased articles are more likely to be considered as fake news.

Moreover, one of the most promising approaches to combat fake news is promoting *news literacy*. Policymakers, educators, librarians, and educational institutions can all help in educating the public—especially younger generations—across all platforms and mediums [14].

Clickbait is a form of link-spam leading to fake content (either news or image). It is a link with a catchy headline that tempts users to click on the link, but it leads to the content entirely unrelated to the headline or less important information. Clickbait works by increasing the curiosity of the user to click the link or image. The purpose of a clickbait is to increase the page views which in turn increase the revenue through ad sense. But when it is used correctly, the publisher can get the readers attention, if not the user might leave the page immediately. Publishers employ various cognitive tricks to make the readers click the links. They write headlines to grab the attention of the readers by provoking their emotions like anger, anxiety, humor, excitement, inspiration, surprise. Another way is by increasing the curiosity of the readers by presenting them with something they know a little bit but not many details about the topic. For example, headlines like "You won't believe what happens next?" provoke the curiosity of the readers and make them click.

Rumors are pieces of information whose veracity is unverifiable and spreads very easily. Their source is unknown, so most of the time the rumors are destructive and misleading. Rumors start as something true and get exaggerated to the point that it is hard to prove. They are often associated with breaking news stories [15]. Kwon et al. [16] report on many interesting findings on rumor spreading dynamics such as (1) a rumor flows from low-degree users to high-degree users, (2) a rumor rarely initiate a conversation and people use speculative words to express doubts about their validity when discussing rumors, and that rumors do not necessarily contain different sentiments than non-rumors. Friggeri et al. [17] analyzed the propagation of known rumors from `Snopes.com` in Facebook and their evolution over time. They found that rumors run deeper in the social network than reshare cascades in general and that when a comment refers to a rumor and contains a link to a Snopes article, then the likelihood that a reshare of a rumor will be deleted increases. Unlike rumors, **hoaxes** consist of false information pretending to be true information and often intended as a joke. Kumar et al. [18] show that 90% of hoaxes articles in Wikipedia are identified in 1 h after their approval, while 1% of hoaxes survive for over 1 year.

Misinformation is also spread through **counterfeit websites** that disguise as legitimate sites. For instance, `ABCnews.com.co` and `Bloomberg.ma` are examples of fake websites. They create more impact and cause severe damage when these sites happen to be subject specific to medical, business, etc.

Also, online videos can contain misinformation. For instance, Youtube videos can have clickbaiting titles, spam in the description, inappropriate or not relevant tags to the videos, etc. [19]. This metadata is used to search and retrieve the video and misinformation in the title or the tags lead to increase the video's views and, consequently, the user' monetization. Sometimes online videos are entirely fake and can be automatically generated via machine learning techniques [20]. As compared to recorded videos, computer-generated ones lack the imperfections, a feature that is hard to incorporate in a machine-learning based algorithm to detect fake videos [21].

3   Detecting Misinformation on the Web

To protect the Web from misinformation, researchers focused on detecting misbehavior, i.e., malicious users such as vandals, spammers, fraudulent reviewers, rumors and fake news spreaders that are responsible for creating and sharing misinformation, or detecting whether or not a given piece of information is false.

In the following, we survey the main methods proposed in the literature to detect either the piece of misinformation or the user causing it. Table 1 summarizes all the related work grouped by misinformation type.

3.1   Vandalism

Plenty of work has been done on detecting vandalism, especially on Wikipedia. One of the first works is the one by Potthast et al. [22] that uses feature extraction (including some linguistic features) and machine learning and validate them on

Table 1 Related work in
detecting misinformation by
type

Types of misinformation	Related work
Vandalism	[22–30]
Spam	[31–44]
Fraudulent reviews	[45–56]
Fake news	[2, 4, 57–67]
Clickbaits	[68–71]
Rumors	[3, 72–80]
Hoaxes	[18, 81]

the PAN-WVC-10 corpus: a set of 32K edits annotated by humans on Amazon
Mechanical Turk [23]. Adler et al. [24] combined and tested a variety of proposed
approaches for vandalism detection including natural language, metadata [25], and
reputation features [26]. Kiesel et al. [27] performed a spatiotemporal analysis of
Wikipedia vandalism revealing that vandalism strongly depends on time, country,
culture, and language. Beyond Wikipedia, vandalism detection has also been
addressed in other platforms such as Wikidata [28] (the Wikimedia knowledge base)
and OpenStreetMaps [29].

Currently, ClueBot NG [7] and STiki [8] are the state-of-the-art tools used
by Wikipedia to detect vandalism. ClueBot NG is a bot based on an artificial
neural network which scores edits and reverts the worst-scoring edits. STiki is an
intelligent routing tool which suggests potential vandalism to humans for definitive
classification. It works by scoring edits by metadata and reverts and computing
a reputation score for each user. Recently, Wikimedia Foundation launched a
new machine learning-based service, called Objective Revision Evaluation Service
(ORES) [82] which measures the level of general damage each edit causes. More
specifically, given an edit, ORES provides three probabilities predicting (1) whether
or not it causes damage, (2) if it was saved in good-faith, and (3) if the edit
will eventually be reverted. These scores are available through the ORES public
API [83].

In our previous work [30], we addressed the problem of vandalism in Wikipedia
from a different perspective. We studied for the *first* time the problem of detecting
vandal users and proposed VEWS, an early warning system to detect vandals before
other Wikipedia bots.[1] Our system leverages differences in the editing behavior of
vandals vs. benign users and detect vandals with an accuracy of over 85% and
outperforms both ClueBot NG and STiki. Moreover, as an early warning system,
VEWS detects, on average, vandals 2.39 edits before ClueBot NG. The combination
of VEWS and Cluebot NG results in a fully automated system that does not leverage
any human input (e.g., edit reversion) and further increases the performances.

Another mechanism used by Wikipedia to protect against content damage is *page
protection*, i.e., placing restrictions on the type of user that can edit the page. To the
best of our knowledge, little research has been done on the topic of page protection

---

[1] Dataset and code are available at http://www.cs.umd.edu/~vs/vews/.

in Wikipedia. Hill and Shaw [84] studied the impact of page protection on user patterns of editing. They also created a dataset (they admit it may not be complete) of protected pages to perform their analysis. There are not currently bots on Wikipedia that can search for pages that may need to be protected. Wikimedia does have a script [85] available in which administrative users can protect a set of pages all at once. However, this program requires that the user supply the pages or the category of pages to be protected and is only intended for protecting a large group of pages at once. There are some bots on Wikipedia that can help with some of the wiki-work that goes along with protecting or removing page protection. This includes adding or removing a template to a page that is marked as protected or no longer marked as protected. These bots can automatically update templates if page protection has expired.

3.2 Spam

Regarding spam detection, various efforts have been made to detect spam users on social networks, mainly by studying their behavior after collecting their profiles through deployed social honeypots [31, 32]. Generally, social networks properties [33, 34], posts content [35, 36], and sentiment analysis [37] have been used to train classifiers for spam users detection.

Regarding spam detection in posted content specifically, researchers mainly concentrated on the problem of predicting whether a link contained in an edit is spam or not. URLs have been analyzed by using blacklists, extracting lexical features and redirecting patterns from them, considering metadata or the content of the landing page, or examining the behavior of who is posting the URL and who is clicking on it [38–41]. Another big challenge is to recognize a short URL as spam or not [42].

Link-spamming has also been studied in the context of Wikipedia. West et al. [43] created the first Wikipedia link-spam corpus, identified Wikipedia's link spam vulnerabilities, and proposed mitigation strategies based on explicit edit approval, refinement of account privileges, and detecting potential spam edits through a machine learning framework. The latter strategy, described by the same authors in [44], relies on features based on (1) article metadata and link/URL properties, (2) HTML landing site analysis, and (3) third-party services used to discern spam landing sites. This tool was implemented as part of STiki (a tool suggesting potential vandalism) and has been used on Wikipedia since 2011. Nowadays, this STiki component is inactive due to a monetary cost for third-party services.

3.3 Rumors and Hoaxes

The majority of the work focused on studying rumors and hoaxes characteristics, and very little work has been done on automatic classification [3, 72, 73]. Qazvinian et al. [74] addressed the problem of rumor detection in Twitter via temporal, content-based and network-based features and additional features extracted from hashtags and URLs present in the tweet. These features are also effective in identifying disinformers, e.g., users who endorse a rumor and further help it to spread. Zubiaga et al. [75] identify whether or not a tweet is a rumor by using the context of from earlier posts associated with a particular event. Wu et al. [76] focused on early detection of emerging rumors by exploiting knowledge learned from historical data. More work has been done for rumor or meme source identification in social networks by defining ad-hoc centrality measures, e.g., rumor centrality, and study rumor propagation via diffusion models, e.g., the SIR model [77–80].

Kumar et al. [18] proposed an approach to detect hoaxes according to article structure and content, hyperlink network properties, and hoaxes' creator reputation. Tacchini et al. [81] proposed a technique to classify Facebook posts as hoaxes or non-hoaxes on the basis of the users who "liked" them.

3.4 Fraudulent Reviews

A Fraudulent review (or deceptive opinion spam) is a review with fictitious opinions which are deliberately written to sound authentic. There are many characteristics that are often hallmarks of fraudulent reviews:

1. *There is no information about the reviewer.* Users who only post a small number of reviews or have no profile information or social connections are more likely to be fibbing.
2. *The opinions are all-or-nothing.* Fabricated reviews tend to be more extreme (all 5 stars or all one star).
3. *Several are posted at once.* Suddenly a product or company with no reviews or one every few months will have five in a row all mentioning something similar, from the same day which indicates that a company paid for a batch of reviews.
4. *They use smaller words.* Scientists say it takes more brainpower to tell a lie than a truth; when we're telling a lie, our vocabulary tends to suffer because we're already expending mental energy on the fabrication. As a result, fraudulent reviews are characterized by shorter words according to research.
5. *They are very short.* Since fraudulent review mills may only pay a few dollars (or less) per review, there's an incentive for a writer to dash them off quickly.

Jindal and Liu [45] first studied deceptive opinion spam problem and trained machine learning-based models using features based on the opinion content, user, and the product itself. Ott et al. [46] created a benchmark dataset by collecting real reviews from TripAdvisor and employing Amazon Mechanical Turk workers to write fraudulent reviews. They got 90% accuracy in detecting fraudulent reviews on their dataset by using psycholinguistic-based features and text-based features (bigrams). However, Mukherjee et al. [47] found out that the method proposed by Ott et al. is not enough to have good performances on a larger and more realistic dataset extracted from Yelp, but behavioral features on the user who wrote the review performed very well (86% accuracy). They also reported that the word distribution between fake and real reviews is very different in the dataset by Ott et al., while this is not true in their more realistic Yelp dataset.

Since then, researchers started focusing more on the problem of detecting opinion spammers (or fraudulent reviewers), rather than fraudulent reviews. Fei et al. [48] discovered that a large number of opinions made use of a sudden burst either caused by the sudden popularity of the product or by a sudden invasion of a large number of fake opinions including some of the features of real users. They used this finding to design an algorithm that applies loopy belief propagation on a network of reviewers appearing in different bursts to detect opinion spammers. Several other works considered features extracted from reviewer behavior as well [49–51].

Rayana and Akoglu [52] also applied loopy belief propagation on a user–product bipartite network with signed edges (positive or negative reviews) and considered metadata (text, timestamp, rating) to assign prior probabilities of users being spammers, reviews being fake, and products being targeted by spammers. Wang et al. [53] proposed three measures (the trustworthiness of the user, the honesty of the review, and the reliability of the store) to be computed on the user-product-store network. Hooi et al. [54] proposed the BirdNest algorithm that detects opinion spammers according to the fact that (1) fraudulent reviews occur in short bursts of time and (2) fraudulent user accounts have skewed rating distributions. Kumar et al. [55] bridged network data and behavioral data to define measures for computing the fairness (or trustworthiness) of the reviewer, the goodness (or quality) of the product, and the reliability of the review. Several other works have been proposed to detect a group of opinion spammers. For instance, CopyCatch [56] leveraged the lockstep behavior, i.e., groups of users acting together, generally liking the same pages at around the same time.

3.5   Fake News

To identify fake news, the majority of the approaches proposed in the literature have focused on machine learning-based approaches working with features extracted from news content and social context [2].

News content-based features include both linguistic features extracted from the text of the news, metadata-based features such as news source (author and/or publisher), headlines, etc., and visual-based features extracted from images and videos associated with the news. For instance, Seyedmehdi and Papalexakis [57] proposed a solution based on extracting latent features from news article text via tensor decomposition to categorize fake news as extreme bias, conspiracy theory, satire, junk science, hate group, or state news. Potthast et al. [58] used the writing style of the articles to identify extremely biased news from the neutral one by using the techniques called unmasking. This model used the news domain specific style features like ratios of quoted words, external links, the average length of the paragraph, etc. Horne and Adali [59] considered both news body and headline for determining the validity of news. They found out that fake and real news have drastically different headlines as they were able to obtain a 0.71 accuracy when considering the number of nouns, lexical redundancy (TTR), word count, and the number of quotes. Further, the study found that fake titles contain different sorts of words (stop words, extremely positive words, and slang among others) than titles of real news articles. Pérez-Rosas et al. [60] analyzed the news body content only and achieved an accuracy up to 0.76 in detecting fake news. They also tested cross domain classification obtaining poor performances by training in one dataset and testing in the other one. Jin et al. [61] used only visual and statistical features extracted from news images for microblogs news verification.

Social context-based features consider (1) the profile and characteristics of users creating and spreading the news (e.g., number of followers/followees, number of posts, credibility and the reliability of the user) also averaged among all the users related to particular news, (2) users' opinion and reactions towards social media posts (post can potentially contain fake news), (3) various type of networks such as friendship networks, co-occurrence networks (network formed based on the number of posts the user write related to the news), or diffusion network where edges between users represent information dissemination paths among them.

Kim et al. [62] propose methods to not only detect the fake news but also to prevent the spread of fake news by making the user flag fake news and used reliable third-parties to fact check the news content. They developed an online algorithm for this purpose, so it works at the time of user spreading the fake news thus preventing it from spreading. Jin et al. [63] developed a method for detecting fake news by using the users' viewpoints to find relationships such as support or oppose and by building a credibility propagation network by using these relationships. Wu and Liu [64] used the way news spread through the social network to find the fake news. They used graph mining method to analyze the social network and recurrent neural networks to represent and classify propagation pathways of a message.

Finally, *hybrid methods* combine the two previous approaches. For instance, Ruchansky et al. [65] used temporal behavior of users and their response and the text content of the news to detect the fake news. They proposed the CSI model (Capture, Score, and Integrate) to classify the news article. Fairbanks et al. [66] show that a content-based model can identify publisher political bias while a structural analysis of web links is enough to detect whether the news is credible or not. Shu et al. [67] exploited both fake news content and the relationship among publishers, news pieces, and users to detect fake news.

Regarding *clickbait detection* specifically, Chakraborty et al. [68] build personalized automatic blocker for clickbait headlines by using a rich set of features that use sentence structure, word patterns, N-gram features, and clickbait language. Their browser extension 'Stop-Clickbait' warns users for potential clickbaited headlines. Potthast et al. [69] used Twitter datasets to identify messages in social media that lead to clickbait. They gathered tweets from various publishers and constructed features based on teaser message, linked web page, and meta information. Anand et al. [70] used three variants of bidirectional RNN models (LSTM, GPU, and standard RNNs) for detecting clickbait headlines They used two different word embedding techniques such as distributed word embeddings and character-level word embeddings. Chen et al. [71] examined a hybrid approach for clickbait detection by using text-based and non-text based clickbaiting cues. While textual cues use text-based semantic and syntactical analysis, non-textual cues relate to image and user behavior analysis.

4   Case Study: Protecting Wikipedia Content Quality

Wikipedia is the world's biggest free encyclopedia read by many users every day. Thanks to the mechanism by which anyone can edit, its content is expanded and kept constantly updated. However, malicious users can take advantage of this open editing mechanism to seriously compromise the quality of Wikipedia articles. As we have seen in Sect. 2, the main form of content damaging in Wikipedia is vandalism, but other types of damaging edits are also common such as page spamming [6] and dissemination of false information, e.g., through hoax articles [86].

In this section, we discuss our research effort to ensure the content integrity of Wikipedia. W start by introducing the DePP system, which is the state-of-the-art tool for detecting article pages to protect [87, 88] in Wikipedia. Page protection is a mechanism used by Wikipedia to place restrictions on the type of users that can make edits to prevent vandalism, libel, or edit wars. Our DePP system achieves an accuracy of 92.1% across multiple languages and significantly improves over baselines.

Then, we present our work on spam users identification [89]. We formulate the problem as a binary classification task and propose a new system, called WiSDe, based on a set of features representing user editing behavior to separate spam users from benign ones. Our results show that WiSDe reaches 80.8% classification accuracy and 0.88 mean average precision and beat ORES, the most recent tool developed by Wikimedia to assign damaging scores to edits.

Finally, we discuss our related work [90, 91] on detecting editors who will stop contributing to the encyclopedia. In fact, one of the main problems of fighting vandalism on Wikipedia is that newcomers are considered suspicious from veteran users who often delete their contributions causing the non-integration of newcomers in the community. We think that the early prediction of inactive users is useful for Wikipedia administrators or other users to perform recovering actions in time to avoid the loss of contributors. This section does not provide new results but collects the ones presented in our prior publications.

4.1   Detecting Pages to Protect

The first problem we address consists of deciding whether or not Wikipedia administrators should protect a page. *Page protection* consists of placing restrictions on the type of users that can edit a Wikipedia page. Examples of protected pages on English Wikipedia are *Drug* and *Biology*. Users can recognize this kind of pages by the image of a lock in the upper right-hand corner of the page. Common motivations that an administrative user may have in protecting a page include (1) consistent vandalism or libel from one or more users, and (2) avoiding edit wars [92]. An edit war is when two users cannot agree on the content of an article and one user repeatedly reverts the other's edits.

There are different levels of page protection for which different levels of users can make edits (or, in general, perform actions on the page): fully protected pages can be edited (or moved) only by administrators, semi-protected pages can be modified only by autoconfirmed users, while move protection does not allow pages to be moved to a new title, except by an administrator. Page protections can also be set for different amounts of time, including 24 or 36 h, or indefinitely.

Currently, the English Wikipedia contains over five million pages. Only a small percentage of those pages are currently protected less than 0.2%. However, around 17 pages become protected every day (according to the number of protected pages from May 6 through Aug 6, 2016). This ratio shows how it is difficult for administrative users to monitor overall Wikipedia pages to determine if any need to be protected. Users can request pages to be protected or unprotected, but an administrative user would have to analyze the page to determine if it should be protected, what level of protection to give, and for how long the protection should last, if not indefinitely. All this work is currently *manually* done by administrators.

To overcome this problem, we propose DePP, the *first* automated tool to detect pages to protect in Wikipedia. DePP is a machine learning-based tool that works with two novel set of features based on (1) *users page revision behavior* and (2) *page categories*. More specifically, the first group of features includes the following six base features:

E1   *Total average time between revisions*: pages that have very few edits over a long period of time are less likely to become protected (as their content is more stable) than pages with many edits that happen with little time between them.

E2   *Total number of users making five or more revisions*: this feature counts the number of users who make more than five edits to a page.

E3   *Total average number of revisions per user*: if there are many users making a few changes to a page, it is less likely to become protected than if a few users are making a lot of changes to a page.

E4   *Total number of revisions by non-registered users*: this feature measures the number of changes made to a page from non-registered users. If a user has not spent the time to set up an account, it is less likely that they are a proficient user and more likely to be a spammer or vandal. Therefore, the more non-registered users that are editing a page, the more likely it is that the page may need to be protected.

E5   *Total number of revisions made from mobile device*: similar to feature E4, this
      feature looks at the number of revisions that are tagged as coming from a
      mobile device. This is a useful feature because users making changes from
      a mobile device are not likely to be sitting down to spend time making
      revisions to a page that would add a lot of value. It is possible that a
      user making a change from a mobile device is only adding non-useful
      information, vandalizing a page, or reverting vandalism that needs to be
      removed immediately.
E6   *Total average size of revisions*: it is possible that users vandalizing a page,
      or adding non-useful information would make an edit that is smaller in size.
      This is opposed to a proficient user who may be adding a large amount of
      new content to a page. For this reason, we measure the average size of an edit.
      Small edits to a page may lead to a page becoming protected more than large
      edits would.

In addition to the above base features, we also include an additional set of
features taking into account the page editing pattern over time. We define these
features by leveraging the features E1–E6 as follows. For each page, we consider
the edits made in the latest 10 weeks and we split this time interval into time frames
of 2 weeks (last 2 weeks, second last 2 weeks, etc.). Then, we compute the base
features E1–E6 within each time frame and compute the standard deviation of each
base features in the 10 weeks time interval. This produces six new features whose
idea is to measure how much features E1–E6 are stable over time. For instance,
for normal pages with solid content, we may observe fewer edits of smaller size
representing small changes in the page, corresponding to a low standard deviation
for features E1, E3, and E6. On the other hand, a higher standard deviation of the
base features can describe a situation where the content of the page was initially
stable, but suddenly we observe a lot of edits from many users, which may indicate
the page is under a vandalism attack and may need protection.

The second group of features use information about page categories and includes:

NC   *Number of categories a page is marked under*;
PC   *Probability of protecting the page given its categories*: given all the pages
      in the training set T and a page category c, we compute the probability
      $pr(c)$ that pages in category c are protected as the percentage of pages in T
      having category c that are protected. Then, given a page p having categories
      $c_1, \ldots, c_n$, we compute this feature as the probability that the page is in at
      least one category whose pages have a high probability to be protected as

$$PC(p) = 1 - \prod_{i=1}^{n}(1 - pr(c_i)).$$

We also define another group of features that shows how much features E1–
E6 vary for a page p w.r.t. the average of these values among all the pages in the
same categories as p. Specifically, given the set of pages in the training set T, we

computed the set C of the top-k most frequent categories. Additionally, for each category $c \in C$, we averaged the features E1–E6 among all the pages (denoted by T_c) having the category c in the training set. Then, for each page p we computed the following set of features, one for each feature Ei ($1 \leq i \leq 6$) and for each category $c \in C$ as follows:

$$
C(Ei, c) = \begin{cases} Ei(p) - \text{avg}_{p' \in T_c}(Ei(p')) & \text{if } p \text{ is in category } c \\ 0 & \text{otherwise} \end{cases}
$$

where $Ei(p)$ is the value of the feature Ei for the page p. The aim of this group of features is to understand if a page is anomalous w.r.t. other pages in the same category. All the features that we propose are language independent as they do not consider page content. As a consequence, DePP is general and able to work on any version of Wikipedia.

To test our DePP system, we built four balanced datasets, one for each of the following Wikipedia versions: English, German, French, and Italian. Each dataset contains all edit protected articles until to Oct. 12, 2016, an almost equal number of randomly selected unprotected pages, and up to the last 500 most recent revisions for each selected page. The sizes of these datasets[2] are reported in Table 2. For protected pages, we only gathered the revisions up until the most recent protection. If there was more than one recent protection, we gathered the revision information between the two protections. This allowed us to focus on the revisions leading up to the most recent page protection. Revision information that we collected included the user who made the revision, the timestamp of the revision, the size of the revision, the categories of the page, and any comments, tags or flags associated with the revision.

The DePP accuracy in the prediction task on 10-fold cross validation is reported for random forest (the best performing algorithm as compared to Logistic Regression, SVM, and K-Nearest Neighbor) in Table 3. As we can see, DePP can classify pages to protect from pages that do not need protection with an accuracy greater

Table 2 English, German, French, and Italian Wikipedia datasets used in to test our DePP system

	English	German	French	Italian
Protected pages	7968	1722	524	171
Unprotected pages	7889	1706	512	168
Number of edits	2.2M	311K	106K	29K

Table 3 DePP accuracy results and comparison with baselines

	English	German	French	Italian
B1+B2+B3	78%	48%	77%	43%
DePP	**95%**	**93%**	**93%**	**91%**

Everything is computed with random forest (best classifier)
Best scores are highlighted in bold

---

[2]Datasets available at http://bit.ly/wiki_depp.

than 91% in all the four languages. As no automated tool detecting which page
to protect exists in Wikipedia, we defined some baselines to compare our results.
One of the main reasons for protecting a page on Wikipedia is to stop edit wars,
vandalism or libel from happening or continuing to happen on a page. Thus, we
used the following baselines:

B1   *Number of revisions tagged as "Possible libel or vandalism"*: These tags
     are added automatically without human interference by checking for certain
     words that might be likely to be vandalism. If a match is found, the tag is
     added.
B2   *Number of revisions that Wikipedia bots or tools reverted as possible vandal-
     ism*: number of reverted edits in the page made by each one of these tools.
     We considered Cluebot NG and STiki for English Wikipedia and Salebot for
     French Wikipedia. We did not find any bot fighting vandalism for German or
     Italian Wikipedia.
B3   *Number of edit wars between two users in the page*: Edit warring occurs when
     two users do not agree on the content of a page or revision. Therefore, we
     count the number of edit wars within the revision history of a page as another
     baseline. In some Wikipedia languages, e.g., German and Italian, there is an
     explicit tag denoting edit wars. For English and French Wikipedia, we define
     an edit war as one user making a revision to a page, followed by another user
     reverting that revision, and this pattern happens 2 or 3 consecutive times.

As we can see in Table 3, `DePP` significantly beats the combination of all the three
baselines across all the languages. By analyzing the most important features, we
found that features E1 (total average time between revisions) and PC (probability
of protecting the page given its categories) consistently appear within the top-15
features in all the four languages considered. Wikipedia editors spend less time in
revising pages that end up being protected. For instance, in English Wikipedia the
mean average time between revisions in 5.8 days for protected pages and 2.9 months
for unprotected ones. Also, a protected page is more likely to be in categories
that have other protected pages than an unprotected page (a probability of 0.84 on
average vs. 0.52 in English Wikipedia).

   In the real-world scenario, we have more unprotected pages than unprotected
ones.[3] Thus, we performed an experiment where we created an unbalanced setting
by randomly selecting pages at a ratio of 10% protected and 90% unprotected (due
to the size of the data we have, we could not reduce this ratio further). Then, we
performed 10-fold cross-validation and measured the performance by using the
area under the ROC curve (AUROC). We used class weighting to deal with class
imbalance. Due to the randomness introduced, we repeated each experiment 10
times and averaged the results. Results are reported in Table 4. We observe that
AUROC values are pretty high across all the dataset considered and outperforms

---

[3]Currently, there is a 0.16% of protected pages in English Wikipedia, 0,09% in German, 0.04% in
French, and 0.015% in Italian.

Table 4 DePP AUROC
results and comparison with
baselines in the unbalanced
setting (10% protected pages,
90% unprotected)

	English	German	French	Italian
B1+B2+B3	0.77	0.50	0.80	0.50
DePP	**0.97**	**0.97**	**0.97**	**0.96**

Everything is computed with random forest (best classifier)
Best scores are highlighted in bold

the baselines. In comparison, AUROC values for the balanced setting are 0.98 for English Wikipedia, 0.98 for German, 0.97 for French, and 0.93 for Italian. Thus, performance does not drop when considering a more real-world unbalanced scenario. Moreover, as shown in [93], AUROC values do not change with changes in the test distribution, thus the above AUROC values for the balanced setting are generalizable to an unbalanced one. Random forest results the best classifier in both the balanced and unbalanced setting.

4.2   Spam Users Identification

Another problem that compromises the content quality of Wikipedia articles is spamming. Currently, no specific tool is available on Wikipedia to identify neither spam edits or spam users. Tools like Cluebot NG and STiki are tailored toward vandalism detection, while ORES is designed to detect damaging edits in general. As in the case of page protection, the majority of the work to protect Wikipedia from spammers is done *manually* by Wikipedia users (patrollers, watchlisters, and readers) who monitor recent changes in the encyclopedia and, eventually, report suspicious spam users to administrators for definitive account blocking. To fight spammers on Wikipedia, we study the problem of identifying spam users from good ones [89]. Our work is closer in spirit to [30] as the aim is to classify users by using their editing behavior instead of classifying a single edit as vandalism [7, 8], spam [44] or generally damaging [82].

We propose a system, called WiSDe (Wikipedia Spammer Detector), that uses a machine learning-based framework with a set of features which are based on research that has been done regarding typical behaviors exhibited by spammers: similarity in edit size and links used in revisions, similar time-sensitive behavior in edits, social involvement of a user in the community through contribution to Wikipedia's talk page system, and chosen username. We did not consider any feature related to edit content so that our system would be language independent and capable of working for all Wikipedia versions. Also, the duration of a user's edit history, from the first edit to her most recent edit, is not taken into account as this feature is biased towards spammers who are short-lived due to being blocked by administrators. Finally, we do not rely on third-party services, so there is no overhead cost as in [44].

The list of features we considered to build `WiSDe` are as follows:

User's Edit Size Based Features

S1   *Average size of edits*—since spammers in Wikipedia are primarily trying to promote themselves (or some organization) and/or attract users to click on various links, the sizes of spammers' edits are likely to exhibit some similarity when compared to that of benign users.

S2   *Standard deviation of edit sizes*—since many spammers make revisions with similar content, the variation in a user's edit sizes is likely not to be very large when compared to benign users.

S3   *Variance significance*—since variance in a spam user's edits can change based on a user's average edit size, normalizing a user's standard deviation of edit sizes by their average edit size may balance any difference found by considering the standard deviation alone.

Editing Time Behavior Based Features

S4   *Average time between edits*—spammers across other social media tend to perform edits in batches and in relatively rapid succession, while benign Wikipedia users dedicate more time in curating the article content and then make edits more slowly than spammers.

S5   *Standard deviation of time between edits*—the consistency in timing of spammers' edits tends to be somewhat mechanical, while benign users tend to edit more sporadically.

Links in Edit Based Features

S6   *Unique link ratio*—since spammers often post the same links in multiple edits, a measure of how unique any links that a user posts may be very useful in helping to determine which users are spammers. This measure is calculated for any user that has posted a minimum of two links in all of their edits, and it is the ratio of unique links posted by a user to the total number of links posted by the user (considering only the domain of the links)

S7   *Link ratio in edits*—since spammers on Wikipedia are known to post links in an effort to attract traffic to other sites the number of edits that a user makes which contain links is likely a useful measure in determining spammers from benign users.

Talk Page Edit Ratio Since talk pages do not face the public and are only presented to a user that specifically clicks on one, spammers are less likely to get very many views on these pages, and, therefore are much less likely to make edits to talk pages. Because of this, the ratio of talk pages edited by a user that correspond with the main article pages that a user edits is considered a possible good indicator of whether a user is a spammer or not. We denote this feature by S8.

Username Based Features Zafarani and Liu [94] showed that aspects of users' usernames themselves contain information that is useful in detecting malicious users. Thus, in addition to the features based on users' edit behaviors, we also considered four additional features related to the user's username itself. These four features are: the *number of digits in a username* (S9), the *ratio of digits in a username* (S10), the *number of leading digits in a username* (S11), and the *unique character ratio in a username* (S12).

To test our `WiSDe` system, we built a new dataset[4] containing 4.2K (half spammer and half benign) users and 75.6K edits as follows. We collected all Wikipedia users (up to Nov. 17, 2016) who were blocked for spamming from two lists maintained on Wikipedia: "Wikipedians who are indefinitely blocked for spamming" [95] and "Wikipedians who are indefinitely blocked for link spamming" [96]. The first list contains all spam users blocked before Mar 12, 2009, while the second one includes all link-spammers after Mar 12, 2009, to today. We gathered a total of 2087 spam users (we only included users who did at least one edit) between the two lists considered.

In order to create a balanced dataset of spam/benign users, we randomly select a sample of benign Wikipedia users of roughly the same size as the spammer user set (2119 users). To ensure these were genuine users, we cross-checked their usernames against the entire list of blocked users provided by Wikipedia [97]. This list contains all users in Wikipedia who have been blocked for any reason, spammers included. For each user in our dataset, we collected up to their 500 most recent edits. For each edit, we gathered the following information: edit content, time-stamp, whether or not the edit is done on a Talk page, and the damaging score provided by ORES.

We run 10-fold cross-validation on several machine learning algorithms, namely SVM, Logistic Regression, K-Nearest Neighbor, Random Forest, and XGBoost, to test the performances of our features. Experimental results are shown in Table 5 for the best performing algorithm (XGBoost). Here we can see that `WiSDe` is able to classify spammers from benign users with 80.8% of accuracy and it is a valuable tool in suggesting potential spammers to Wikipedia administrators for further investigation as proved by a mean average precision of 0.88.

Feature importance analysis revealed that the top three most important features for spammers identifications are: *Link ratio in edits*, *Average size of edits*, and *Standard deviation of time between edits*. As expected, spammers use more links in their edits. The average value of this feature is 0.49 for spammers and 0.251 for benign users. Also, benign users put more diverse links in their revisions than spammers (0.64 vs. 0.44 on average). We also have that spammer's edit size is

Table 5 `WiSDe` spammers identification accuracy and Mean Average Precision (MAP) results in comparison with ORES

	Accuracy	MAP
ORES	69.7%	0.695
`WiSDe`	**80.8%**	**0.880**
`WiSDe + ORES`	82.1%	**0.886**

Everything is computed with XGBoost
Best scores are highlighted in bold

---

[4]Dataset available at http://bit.ly/wiki_spammers.

smaller, and they edit faster than benign users. Regarding edits on talk pages, we have that the majority of the users are not using talk pages (percentage for both benign users and spammers is 69.7%). However, surprisingly, we have that, among users editing talk pages, the talk page edit ratio is higher for spammers (0.2) than for benign users (0.081), and we observe a group of around 303 spammers trying to gain visibility by making numerous edits on talk pages. Finally, username based features contribute to an increase in accuracy prediction by 2.9% (from 77.9% to 80.8%) and Mean Average Precision by 0.019 (from 0.861 to 0.880).

We compared WiSDe with ORES only, as the tool proposed in [44] is no longer used, and Cluebot NG and STiki are explicitly designed for vandalism and not spam. To compare our system with ORES, we considered the edit damaging score. More specifically, given a user and all her edits, we computed both the average and maximum damaging score provided by ORES and used these as features for classification. Results on 10-fold cross-validation with XGBoost (the best performing classifier) are reported in Table 5, as well. As we can see, ORES performances are poor for the task of spammer detection (69.7% of accuracy and mean average precision of 0.695). However, combining our features with ORES further increases the accuracy to 82.1%.

In reality, spam users are greatly outnumbered by benign users. Thus, similarly to what we did in the previous section, we also created an unbalanced dataset to test our system WiSDe by randomly selecting users at a ratio of 10% spammers and 90% of benign users. Then, we performed 10-fold cross-validation and measured the performance by using the area under the ROC curve (AUROC). To deal with class imbalance, we oversampled the minority class in each training set by using SMOTE [98]. We also considered class weighting, but we found that SMOTE is performing the best. Due to the randomness introduced, we repeated each experiment 10 times and averaged the results. Table 6 reports the results for this experiment. As we can see, even with class imbalance, WiSDe reaches a good AUROC of 0.842 (in comparison we have an AUROC of 0.891 for the balanced setting) and significantly improve over ORES (AUROC of 0.736). However, adding ORES features to ours helps to increase the AUROC to 0.864.

Table 6 WiSDe vs. ORES performance in the unbalanced setting

	AUROC
ORES	0.736
WiSDe	**0.842**
WiSDe + ORES	**0.864**

Everything is computed by using XGBoost

Best scores are highlighted in bold

4.3 Content Quality Protection and User Retention

As we have seen in this chapter, a lot of research has been done with the aim of maintaining the trustworthiness, legitimacy, and integrity of Wikipedia content. However, one big drawback of deploying anti-vandalism and anti-spam tools is that veteran editors started to suspiciously look at newcomers as potential vandals and rapidly and unexpectedly deleted contributions even from good-faith editors. Many newcomers, in fact, face social barriers [99] preventing them from the integration in the editor community, with the consequence of stop editing after a certain period of time [100, 101]. As Halfaker et al. [102] have pointed out, Wikipedia is not anymore the encyclopedia that anyone can edit but rather *"the encyclopedia that anyone who understands the norms, socializes himself or herself, dodges the impersonal wall of semi-automated rejection, and still wants to voluntarily contribute his or her time and energy can edit."*

The loss of active contributors from any user-generated content community may affect the quantity and quality of content provision not only on the specific community but also on the Web in general. Most importantly, UGC communities mainly survive thanks to the continued participation of their active users who contribute with their content production.

Thus, being able to early predict whether or not a user will become inactive is very valuable for Wikipedia and any other user-generated content community to perform engaging actions on time to keep these users contributing longer. In our related work [90, 91], we addressed the problem of early predict whether or not a Wikipedia editor will become inactive and stop contributing and proposed a predictive model based on users' editing behavior that achieves an AUROC of 0.98 and a precision of 0.99 in predicting inactive users. By comparing the editing behavior of active vs. inactive users, we discovered that active users are more involved in edit wars and positively accept critiques, and edit much more different categories of pages. On the other hand, inactive users have more edits reverted and edit more meta-pages (and in particular *User* pages).

Regarding specific actions for engaging editors, the Wikipedia community considers several steps that can be taken to increase the retention rate such as (1) survey newly registered users to capture user's interests and use them for making relevant editing recommendations and (2) connect editors with similar interests to form meaningful contribution teams. They also developed and deployed a tool, called Snuggle [103], to support newcomers socialization.

5   Conclusions

In this chapter, we discussed several types of misinformation that these days exist on the Web such as vandalism, spam, fraudulent reviews, fake news, etc. and provided a survey on how to detect them.

Then, we focused on the specific case study of protecting Wikipedia from misinformation and presented our research on detecting pages to protect and identifying spam users. Our experimental results show that we are able to classify (1) article pages to protect with an accuracy of 92% across multiple languages and (2) spammers from benign users with 80.8% of accuracy and 0.88 mean average precision. Both the methods proposed do not look at edit content and, as a consequence, they are generally applicable to all versions of Wikipedia, not only the English one.

Finally, we discussed a possible solution for newcomers retention, given that many new users do not keep contributing to the encyclopedia because of the tools deployed to fight vandalism.

References

1. L. Wu, F. Morstatter, X. Hu, H. Liu, Mining misinformation in social media, in *Big Data in Complex and Social Networks* (2016), pp. 123–152
2. K. Shu, A. Sliva, S. Wang, J. Tang, H. Liu, Fake news detection on social media: a data mining perspective. ACM SIGKDD Explor. Newslett. **19**(1), 22–36 (2017)
3. A. Zubiaga, A. Aker, K. Bontcheva, M. Liakata, R. Procter, Detection and resolution of rumours in social media: a survey. ACM Comput. Surv. (CSUR) **51**(2), 32 (2018)
4. S. Kumar, N. Shah, False information on web and social media: a survey (2018). arXiv preprint:1804.08559
5. Vandalism in Wikipedia. http://en.wikipedia.org/wiki/Wikipedia:Vandalism
6. Spam in Wikipedia. http://en.wikipedia.org/wiki/Wikipedia:Spam
7. Cluebot_NG. http://bit.ly/ClueBotNG
8. STiki. http://bit.ly/STiki_tool
9. A. Shrestha, F. Spezzano, M.S. Pera, Who is really affected by fraudulent reviews? an analysis of shilling attacks on recommender systems in real-world scenarios, in *Late-Breaking Results track part of the Twelfth ACM Conference on Recommender Systems (RecSys'18)* (2018)
10. Pew Research Center. http://www.journalism.org/2016/12/15/many-americans-believe-fake-news-is-sowing-confusion/
11. S. Vosoughi, D. Roy, S. Aral, The spread of true and false news online. *Science*, **359**(6380), 1146–1151 (2018)
12. https://www.factcheck.org/2016/11/how-to-spot-fake-news/
13. H. Allcott, M. Gentzkow, Social media and fake news in the 2016 election. J. Econ. Perspect. **31**(2), 211–36 (2017)
14. P. America, *Faking News: Fraudulent News and the Fight for Truth* (2018). https://pen.org/faking-news/
15. A. Zubiaga, E. Kochkina, M. Liakata, R. Procter, M. Lukasik, Stance classification in rumours as a sequential task exploiting the tree structure of social media conversations, in *COLING 2016, 26th International Conference on Computational Linguistics, Proceedings of the Conference: Technical Papers, December 11–16, 2016, Osaka, Japan* (2016), pp. 2438–2448
16. S. Kwon, M. Cha, K. Jung, W. Chen, Y. Wang, Aspects of rumor spreading on a microblog network, in *Proceedings of Social Informatics—5th International Conference, SocInfo 2013, Kyoto, Japan, November 25–27, 2013* (2013), pp. 299–308
17. A. Friggeri, L.A. Adamic, D. Eckles, J. Cheng, Rumor cascades, in *Proceedings of the Eighth International Conference on Weblogs and Social Media, ICWSM 2014, Ann Arbor, Michigan, USA, June 1–4, 2014* (2014)

18. S. Kumar, R. West, J. Leskovec, Disinformation on the web: impact, characteristics, and detection of wikipedia hoaxes, in *Proceedings of the 25th International Conference on World Wide Web, WWW 2016, Montreal, Canada, April 11–15, 2016* (2016), pp. 591–602

19. P. Bajaj, M. Kavidayal, P. Srivastava, M.N. Akhtar, P. Kumaraguru, Disinformation in multimedia annotation: misleading metadata detection on youtube, in *Proceedings of the 2016 ACM Workshop on Vision and Language Integration Meets Multimedia Fusion, iVandL-MM@MM 2016, Amsterdam, Netherlands, October 16, 2016* (2016), pp. 53–61

20. A. Nguyen, J. Clune, Y. Bengio, A. Dosovitskiy, J. Yosinski, Plug and play generative networks: conditional iterative generation of images in latent space, in *2017 IEEE Conference on Computer Vision and Pattern Recognition, CVPR 2017, Honolulu, HI, USA, July 21–26, 2017* (2017), pp. 3510–3520

21. E. Gibney, The scientist who spots fake videos. *Nature* (2017)

22. M. Potthast, B. Stein, R. Gerling, Automatic vandalism detection in wikipedia, in *Proceedings of Advances in Information Retrieval, 30th European Conference on IR Research, ECIR 2008, Glasgow, UK, March 30-April 3, 2008* (2008), pp. 663–668

23. M. Potthast, B. Stein, T. Holfeld, Overview of the 1st international competition on wikipedia vandalism detection, in *CLEF 2010 LABs and Workshops, Notebook Papers, 22–23 September 2010, Padua, Italy* (2010)

24. B.T. Adler, L. de Alfaro, S.M. Mola-Velasco, P. Rosso, A.G. West, Wikipedia vandalism detection: combining natural language, metadata, and reputation features, in *International Conference on Intelligent Text Processing and Computational Linguistics (CICLing)* (2011), pp. 277–288

25. A.G. West, S. Kannan, I. Lee, Detecting wikipedia vandalism via spatio-temporal analysis of revision metadata? in *Proceedings of the Third European Workshop on System Security, EUROSEC 2010, Paris, France, April 13, 2010* (2010), pp. 22–28

26. B.T. Adler, L. de Alfaro, I. Pye, Detecting wikipedia vandalism using wikitrust—lab report for PAN at CLEF 2010, in *CLEF 2010 LABs and Workshops, Notebook Papers, 22–23 September 2010, Padua, Italy* (2010)

27. J. Kiesel, M. Potthast, M. Hagen, B. Stein, Spatio-temporal analysis of reverted wikipedia edits, in *Proceedings of the Eleventh International Conference on Web and Social Media, ICWSM 2017, Montréal, Québec, Canada, May 15–18, 2017* (2017), pp. 122–131

28. S. Heindorf, M. Potthast, B. Stein, G. Engels, Vandalism detection in wikidata, in *Proceedings of the 25th ACM International Conference on Information and Knowledge Management, CIKM 2016, Indianapolis, IN, USA, October 24–28, 2016* (2016), pp. 327–336

29. P. Neis, M. Goetz, A. Zipf, Towards automatic vandalism detection in openstreetmap. ISPRS Int. J. Geo Inf. **1**(3), 315–332 (2012)

30. S. Kumar, F. Spezzano, V.S. Subrahmanian, VEWS: a wikipedia vandal early warning system, in *Proceedings of the 21th ACM SIGKDD International Conference on Knowledge Discovery and Data Mining, Sydney, NSW, Australia, August 10–13, 2015* (2015), pp. 607–616

31. G. Stringhini, C. Kruegel, G. Vigna, Detecting spammers on social networks, in *Proceedings of the 26th Annual Computer Security Applications Conference* (2010), pp. 1–9

32. K. Lee, J. Caverlee, S. Webb, Uncovering social spammers: social honeypots + machine learning, in *Proceedings of the 33rd International ACM SIGIR Conference on Research and Development in Information Retrieval* (2010), pp. 435–442

33. J. Song, S. Lee, J. Kim, Spam filtering in twitter using sender-receiver relationship, in *International Workshop on Recent Advances in Intrusion Detection* (2011), pp. 301–317

34. C. Yang, R.C. Harkreader, G. Gu, Die free or live hard? empirical evaluation and new design for fighting evolving twitter spammers, in *International Workshop on Recent Advances in Intrusion Detection*, pp. 318–337 (2011)

35. C. Grier, K. Thomas, V. Paxson, M. Zhang, @ spam: the underground on 140 characters or less, in *Proceedings of the 17th ACM Conference on Computer and Communications Security (CCS)* (2010), pp. 27–37

36. L. Wu, X. Hu, F. Morstatter, H. Liu, Detecting camouflaged content polluters, in *Proceedings of the Eleventh International Conference on Web and Social Media, ICWSM 2017, Montréal, Québec, Canada, May 15–18, 2017* (2017), pp. 696–699
37. X. Hu, J. Tang, H. Gao, H. Liu, Social spammer detection with sentiment information, in *2014 IEEE International Conference on Data Mining (ICDM)* (2014), pp. 180–189
38. S. Lee, J. Kim, Warningbird: detecting suspicious urls in twitter stream. in *19th Annual Network and Distributed System Security Symposium, NDSS 2012, San Diego, California, USA, February 5–8, 2012* (2012)
39. J. Ma, L.K. Saul, S. Savage, G.M. Voelker, Beyond blacklists: learning to detect malicious web sites from suspicious urls, in *Proceedings of the 15th ACM SIGKDD International Conference on Knowledge Discovery and Data Mining, Paris, France, June 28–July 1, 2009* (2009), pp. 1245–1254
40. D.K. McGrath, M. Gupta, Behind phishing: an examination of phisher modi operandi, in *Proceedings of First USENIX Workshop on Large-Scale Exploits and Emergent Threats, LEET '08, San Francisco, CA, USA, April 15, 2008* (2008)
41. C. Cao, J. Caverlee, Detecting spam urls in social media via behavioral analysis, in *Proceedings of Advances in Information Retrieval—37th European Conference on IR Research, ECIR 2015, Vienna, Austria March 29–April 2, 2015* (2015), pp. 703–714
42. D. Antoniades, I. Polakis, G. Kontaxis, E. Athanasopoulos, S. Ioannidis, E.P. Markatos, T. Karagiannis, we.b: the web of short URLs, in *Proceedings of the 20th International Conference on World Wide Web, WWW 2011, Hyderabad, India, March 28–April 1, 2011* (2011), pp. 715–724
43. A.G. West, J. Chang, K. Venkatasubramanian, O. Sokolsky, I. Lee, Link spamming wikipedia for profit, in *CEAS* (2011), pp. 152–161
44. A.G. West, A. Agrawal, P. Baker, B. Exline, I. Lee, Autonomous link spam detection in purely collaborative environments, in *WikiSym* (2011), pp. 91–100
45. N. Jindal, B. Liu, Opinion spam and analysis, in *Proceedings of the International Conference on Web Search and Web Data Mining, WSDM 2008, Palo Alto, California, USA, February 11–12, 2008* (2008), pp. 219–230
46. M. Ott, Y. Choi, C. Cardie, J.T. Hancock, Finding deceptive opinion spam by any stretch of the imagination, in *Proceedings of the 49th Annual Meeting of the Association for Computational Linguistics: Human Language Technologies, 19–24 June, 2011, Portland, Oregon, USA* (2011), pp. 309–319
47. A. Mukherjee, V. Venkataraman, B. Liu, N.S. Glance, What yelp fake review filter might be doing? in *Proceedings of the Seventh International Conference on Weblogs and Social Media, ICWSM 2013, Cambridge, Massachusetts, USA, July 8–11, 2013* (2013)
48. G. Fei, A. Mukherjee, B. Liu, M. Hsu, M. Castellanos, R. Ghosh, Exploiting burstiness in reviews for review spammer detection, in *Proceedings of the Seventh International Conference on Weblogs and Social Media, ICWSM 2013, Cambridge, Massachusetts, USA, July 8–11, 2013* (2013)
49. E. Lim, V.A. Nguyen, N. Jindal, B. Liu, H.W. Lauw, Detecting product review spammers using rating behaviors, in *Proceedings of the 19th ACM Conference on Information and Knowledge Management, CIKM 2010, Toronto, Ontario, Canada, October 26–30, 2010* (2010), pp. 939–948
50. A. Mukherjee, A. Kumar, B. Liu, J. Wang, M. Hsu, M. Castellanos, R. Ghosh, Spotting opinion spammers using behavioral footprints, in *Proceedings of the 19th ACM SIGKDD International Conference on Knowledge Discovery and Data Mining, KDD 2013, Chicago, IL, USA, August 11–14, 2013* (2013), pp. 632–640
51. K.C. Santosh, A. Mukherjee, On the temporal dynamics of opinion spamming: case studies on yelp, in *Proceedings of the 25th International Conference on World Wide Web, WWW 2016, Montreal, Canada, April 11–15, 2016* (2016), pp. 369–379

52. S. Rayana, L. Akoglu, Collective opinion spam detection: Bridging review networks and metadata, in *Proceedings of the 21th ACM SIGKDD International Conference on Knowledge Discovery and Data Mining, Sydney, NSW, Australia, August 10–13, 2015* (2015), pp. 985–994

53. G. Wang, S. Xie, B. Liu, P.S. Yu, Review graph based online store review spammer detection, in *Proceedings of the 11th IEEE International Conference on Data Mining, ICDM 2011, Vancouver, BC, Canada, December 11–14, 2011* (2011), pp. 1242–1247

54. B. Hooi, N. Shah, A. Beutel, S. Günnemann, L. Akoglu, M. Kumar, D. Makhija, C. Faloutsos, BIRDNEST: bayesian inference for ratings-fraud detection, in *Proceedings of the 2016 SIAM International Conference on Data Mining, Miami, Florida, USA, May 5–7, 2016* (2016), pp. 495–503

55. S. Kumar, B. Hooi, D. Makhija, M. Kumar, C. Faloutsos, V.S. Subrahmanian, REV2: fraudulent user prediction in rating platforms, in *Proceedings of the Eleventh ACM International Conference on Web Search and Data Mining, WSDM 2018, Marina Del Rey, CA, USA, February 5–9, 2018* (2018), pp. 333–341

56. A. Beutel, W. Xu, V. Guruswami, C. Palow, C. Faloutsos, Copycatch: stopping group attacks by spotting lockstep behavior in social networks, in *Proceedings of the 22nd International World Wide Web Conference, WWW '13, Rio de Janeiro, Brazil, May 13–17, 2013* (2013), pp. 19–130

57. S. Hosseinimotlagh, E.E. Papalexakis, Unsupervised content-based identification of fake news articles with tensor decomposition ensembles, in *MIS2: Misinformation and Misbehavior Mining on the Web Workshop held in conjunction with WSDM 2018 Feb 9, 2018—Los Angeles, California, USA, 2018* (2018)

58. M. Potthast, J. Kiesel, K. Reinartz, J. Bevendorff, B. Stein, A stylometric inquiry into hyperpartisan and fake news. CoRR, abs/1702.05638 (2017)

59. B.D. Horne, S. Adali, This just in: fake news packs a lot in title, uses simpler, repetitive content in text body, more similar to satire than real news (2017). arXiv preprint:1703.09398

60. V. Pérez-Rosas, B. Kleinberg, A. Lefevre, R. Mihalcea, Automatic detection of fake news, in *Proceedings of the 27th International Conference on Computational Linguistics* (2018), pp. 3391–3401

61. Z. Jin, J. Cao, Y. Zhang, J. Zhou, Q. Tian, Novel visual and statistical image features for microblogs news verification. IEEE Trans. Multimedia **19**(3), 598–608 (2017)

62. J. Kim, B. Tabibian, A. Oh, B. Schölkopf, M. Gomez-Rodriguez, Leveraging the crowd to detect and reduce the spread of fake news and misinformation, in *Proceedings of the Eleventh ACM International Conference on Web Search and Data Mining, WSDM 2018, Marina Del Rey, CA, USA, February 5–9, 2018* (2018), pp. 324–332

63. Z. Jin, J. Cao, Y. Zhang, J. Luo, News verification by exploiting conflicting social viewpoints in microblogs, in *Proceedings of the Thirtieth AAAI Conference on Artificial Intelligence, February 12–17, 2016, Phoenix, Arizona, USA* (2016), pp. 2972–2978

64. L. Wu, H. Liu, Tracing fake-news footprints: characterizing social media messages by how they propagate, in *Proceedings of the Eleventh ACM International Conference on Web Search and Data Mining, WSDM 2018, Marina Del Rey, CA, USA, February 5–9, 2018* (2018), pp. 637–645

65. N. Ruchansky, S. Seo, Y. Liu, CSI: a hybrid deep model for fake news detection. in *Proceedings of the 2017 ACM on Conference on Information and Knowledge Management, CIKM 2017, Singapore, November 06–10, 2017* (2017), pp. 797–806

66. N. Knauf, J. Fairbanks, N. Fitch, E. Briscoe, Credibility assessment in the news: do we need to read? in *MIS2: Misinformation and Misbehavior Mining on the Web Workshop held in conjunction with WSDM 2018 Feb 9, 2018—Los Angeles, California, USA, 2018* (2018)

67. K. Shu, S. Wang, H. Liu, Beyond news contents: the role of social context for fake news detection, in *Proceedings of the Twelfth ACM International Conference on Web Search and Data Mining, WSDM 2019, Melbourne, VIC, Australia, February 11–15, 2019* (2019), pp. 312–320

68. A. Chakraborty, B. Paranjape, S. Kakarla, N. Ganguly, Stop clickbait: detecting and preventing clickbaits in online news media, in *2016 IEEE/ACM International Conference on Advances in Social Networks Analysis and Mining, ASONAM 2016, San Francisco, CA, USA, August 18–21, 2016* (2016), pp. 9–16
69. M. Potthast, S. Köpsel, B. Stein, M. Hagen, Clickbait detection, in *Proceedings of Advances in Information Retrieval—38th European Conference on IR Research, ECIR 2016, Padua, Italy, March 20–23, 2016* (2016), pp. 810–817
70. A. Anand, T. Chakraborty, N. Park, We used neural networks to detect clickbaits: you won't believe what happened next!, in *Proceedings of Advances in Information Retrieval—39th European Conference on IR Research, ECIR 2017, Aberdeen, UK, April 8–13, 2017* (2017) pp. 541–547
71. Y. Chen, N.J. Conroy, V.L. Rubin, Misleading online content: recognizing clickbait as "false news", in *Proceedings of the 2015 ACM Workshop on Multimodal Deception Detection, WMDD@ICMI 2015, Seattle, Washington, USA, November 13, 2015* (2015), pp. 15–19
72. S. Hamidian, M. Diab, Rumor detection and classification for twitter data, in *Proceedings of the Fifth International Conference on Social Media Technologies, Communication, and Informatics (SOTICS)* (2015), pp. 71–77
73. S. Hamidian, M. Diab, Rumor identification and belief investigation on twitter, in *Proceedings of the 7th Workshop on Computational Approaches to Subjectivity, Sentiment and Social Media Analysis* (2016), pp. 3–8
74. V. Qazvinian, E. Rosengren, D.R. Radev, Q. Mei, Rumor has it: identifying misinformation in microblogs, in *Proceedings of the 2011 Conference on Empirical Methods in Natural Language Processing, EMNLP 2011, 27–31 July 2011, John McIntyre Conference Centre, Edinburgh, UK, A meeting of SIGDAT, a Special Interest Group of the ACL* (2011), pp. 1589–1599
75. A. Zubiaga, M. Liakata, R. Procter, Exploiting context for rumour detection in social media, in *International Conference on Social Informatics* (Springer, Berlin, 2017), pp. 109–123
76. L. Wu, J. Li, X. Hu, H. Liu, Gleaning wisdom from the past: early detection of emerging rumors in social media, in *Proceedings of the 2017 SIAM International Conference on Data Mining, Houston, Texas, USA, April 27–29, 2017* (2017), pp. 99–107
77. D. Shah, T. Zaman, Rumors in a network: who's the culprit? IEEE Trans. Infor. Theory **57**(8), 5163–5181 (2011)
78. W. Luo, W.-P. Tay, Finding an infection source under the SIS model, in *IEEE International Conference on Acoustics, Speech and Signal Processing, ICASSP 2013, Vancouver, BC, Canada, May 26–31, 2013* (2013), pp. 2930–2934
79. W. Dong, W. Zhang, C.W. Tan, Rooting out the rumor culprit from suspects, in *Proceedings of the 2013 IEEE International Symposium on Information Theory, Istanbul, Turkey, July 7–12, 2013* (2013), pp. 2671–2675
80. C. Kang, S. Kraus, C. Molinaro, F. Spezzano, V.S. Subrahmanian, Diffusion centrality: a paradigm to maximize spread in social networks. Artif. Intell. **239**, 70–96 (2016)
81. E. Tacchini, G. Ballarin, M.L. Della Vedova, S. Moret, L. de Alfaro, Some like it hoax: automated fake news detection in social networks. CoRR, abs/1704.07506 (2017)
82. ORES. http://bit.ly/wikipedia_ores
83. ORES API. http://ores.wikimedia.org
84. B.M. Hill, A.D. Shaw, Page protection: another missing dimension of wikipedia research, in *Proceedings of the 11th International Symposium on Open Collaboration, San Francisco, CA, USA, August 19–21, 2015* (2015), pp. 15:1–15:4
85. Pywikibot. https://www.mediawiki.org/wiki/Manual:Pywikibot/protect.py
86. Hoaxes on Wikipedia. https://en.wikipedia.org/wiki/Wikipedia:List_of_hoaxes_on_Wikipedia
87. K. Suyehira, F. Spezzano, Depp: a system for detecting pages to protect in wikipedia, in *Proceedings of the 25th ACM International Conference on Information and Knowledge Management, CIKM 2016, Indianapolis, IN, USA, October 24–28, 2016* (2016), pp. 2081–2084

88. F. Spezzano, K. Suyehira, L.A. Gundala, Detecting pages to protect in wikipedia across multiple languages. Soc. Netw. Anal. Min. **9**(1), 10 (2018)
89. T. Green, F. Spezzano, Spam users identification in wikipedia via editing behavior, in *Proceedings of the Eleventh International Conference on Web and Social Media, ICWSM 2017, Montréal, Québec, Canada, May 15–18, 2017* (2017), pp. 532–535
90. H. Arelli, F. Spezzano, Who will stop contributing?: predicting inactive editors in wikipedia, in *Proceedings of the 2017 IEEE/ACM International Conference on Advances in Social Networks Analysis and Mining 2017, Sydney, Australia, July 31–August 03, 2017* (2017), pp. 355–358
91. H. Arelli, F. Spezzano, A. Shrestha, Editing behavior analysis for predicting active and inactive users in wikipedia, in *Influence and Behavior Analysis in Social Networks and Social Media* (2019), pp. 127–147
92. Edit War in Wikipedia. http://en.wikipedia.org/wiki/Wikipedia:Editwarring
93. T. Fawcett, An introduction to ROC analysis. Pattern Recogn. Lett. **27**(8), 861–874 (2006)
94. R. Zafarani, H. Liu, 10 bits of surprise: detecting malicious users with minimum information, in *CIKM* (2015), pp. 423–431
95. http://en.wikipedia.org/wiki/Category:Wikipedians_who_are_indefinitely_blocked_for_spamming
96. http://en.wikipedia.org/wiki/Category:Wikipedians_who_are_indefinitely_blocked_for_link-spamming
97. http://en.wikipedia.org/wiki/Special:BlockList
98. N.V. Chawla, K.W. Bowyer, L.O. Hall, W.P. Kegelmeyer, Smote: synthetic minority over-sampling technique. J. Artif. Intell. Res. **16**, 321–357 (2002)
99. I. Steinmacher, T. Conte, M.A. Gerosa, D.F. Redmiles, Social barriers faced by newcomers placing their first contribution in open source software projects, in *Proceedings of the 18th ACM Conference on Computer Supported Cooperative Work and Social Computing, CSCW 2015, Vancouver, BC, Canada, March 14–18, 2015* (2015), pp. 1379–1392
100. L. Jian, J.K. MacKie-Mason, Why leave wikipedia? in *iConference* (2008)
101. S. Asadi, S. Ghafghazi, H.R. Jamali, Motivating and discouraging factors for wikipedians: the case study of persian wikipedia. Libr. Rev. **62**(4/5), 237–252 (2013)
102. A. Halfaker, R.S. Geiger, J.T. Morgan, J. Riedl, The rise and decline of an open collaboration system: how wikipedia's reaction to popularity is causing its decline. Am. Behav. Sci. **57**(5), 664–688 (2013)
103. A. Halfaker, R.S. Geiger, L.G. Terveen, Snuggle: designing for efficient socialization and ideological critique, in *CHI Conference on Human Factors in Computing Systems, CHI'14, Toronto, ON, Canada—April 26–May 01, 2014* (2014), pp. 311–320

Studying the Weaponization of Social Media: Case Studies of Anti-NATO Disinformation Campaigns

Katrin Galeano, Rick Galeano, Samer Al-Khateeb, and Nitin Agarwal

Abstract Social media provides a fertile ground for any user to find or share information about various events with others. At the same time, social media is not always used for benign purposes. With the availability of inexpensive and ubiquitous mass communication tools, disseminating false information and propaganda is both convenient and effective. In this research, we studied Online Deviant Groups (ODGs) that conduct cyber propaganda campaigns in order to achieve strategic and political goals, influence mass thinking, and steer behaviors or perspectives about an event. We provide case studies in which various disinformation and propaganda swamped social media during two NATO exercises in 2015. We demonstrate ODGs' capability to spread anti-NATO propaganda using a highly sophisticated and well-coordinated social media campaign. In particular, blogs were used as virtual spaces where narratives are framed. And, to generate discourse, web traffic was driven to these virtual spaces via other social media platforms such as Twitter, Facebook, and VKontakte. By further examining the information flows within the social media networks, we identify sources of mis/disinformation and their reach, i.e., how far and how quickly the mis/disinformation could travel and consequently detect manipulation. The chapter presents an in-depth examination of the information networks using social network analysis (SNA) and social cyber forensics (SCF) based methodologies to identify prominent information brokers, leading coordinators, and information competitors who seek to further their own agenda. Through SCF tools, e.g., Maltego, we extract metadata associated with disinformation-riddled websites. The extracted metadata helps in uncovering the implicit relations among various ODGs. We further collected the social network of various ODGs (i.e., their friends and followers) and their communication network (i.e., network depicting the flow of information such as tweets, retweets, mentions,

K. Galeano · R. Galeano · N. Agarwal (✉)
Department of Information Science, University of Arkansas at Little Rock, Little Rock, AR, USA
e-mail: kkaniagalea@ualr.edu; ragaleano@ualr.edu; nxagarwal@ualr.edu

S. Al-Khateeb
Department of Journalism, Media, and Computing, Creighton University, Omaha, NE, USA
e-mail: sameral-khateeb1@creighton.edu

© Springer Nature Switzerland AG 2020   29
M. A. Tayebi et al. (eds.), *Open Source Intelligence and Cyber Crime*, Lecture Notes in Social Networks, https://doi.org/10.1007/978-3-030-41251-7_2

and hyperlinks). SNA helped us identify influential users and powerful groups responsible for coordinating the various disinformation campaigns. One of the key research findings is the vitality of the link between blogs and other social media platforms to examine disinformation campaigns.

Keywords Social media · Weaponization · Cyber forensics · Social network analysis · NATO · Disinformation

1 Introduction

Global communication has accelerated through the use of social media platforms over the past decade and in turn, this social media craze has affected demographics across the globe. An observation made in 2016 while living in The Netherlands, was that of school-age children pedaling their bicycles through busy city streets and staring at the screens of their phones. It even appeared that they were responding to messages while riding their bikes. Just a few years ago, this would have been unheard of, but it is ever present just driving down the street in any city now to see people driving vehicles and texting, posting updates, or even surfing the web.

The youth of yesteryear are the savvy technology operators of today. The kids that grew up playing the Atari 2600 or the Commodore 64 systems have revolutionized the way people communicate in our general day to day lifestyles; can you imagine how our communications cycle will look twenty-five years from now with the millennials that text and ride their bikes at the same time! For example, prior to the home video game industry, electronic gaming was more of a social setting that happened via coin-operated machines at an arcade or the pinball machine at the bar. "Atari bridged the gap, they moved video games from these places to the home. In so doing, they caused a market shift. [They realized] if you're only selling games per play to people in bars, then you're missing out on a whole marketplace of families and kids," Bogost said [1].

Transferring from the mindset of the 'home-based' video gaming world into the modern age of social media communications, look at the CEO of Facebook, Mark Zuckerberg. Facebook had its roots with Atari systems. At the age of twelve, Mr. Zuckerberg created "Zucknet" that was used in his father's dental office. Zucknet was a social messaging network designed to share data on patients and inform hygienists that patients were in the waiting room; essentially "Zucknet" was the grandfather of Facebook [2]. Mindful that not all social media platforms have a kinship to the home video game industry, they do have the universal notion of providing social conversations, sharing, and a gathering point for online business and personal communications. All too often, social media is used as the medium for the change agent. In this case, it does not necessarily represent an individual person but rather a larger concept. The change agent seeks to control the narrative.

Controlling the narrative; irrelevant to the platform, dates back to the 3rd century B.C. at the Platonic Academy in Greece. Aristotle's approach with his three appeals about the general means of persuasion: (1) Logos (logic/reason/proof) (2) Ethos (credibility/trust), and (3) Pathos (emotions/values) have morphed into modern day disinformation campaigns via digital platforms. Controlling the narrative via mass influx of messaging into the information environment or simply being the first to input information—EVEN IF IT IS FALSE—has been able to gain momentum expeditiously.

This has developed into controlling the narrative via social communications. Social communication happens online, at a business meeting and even in class-rooms. Influencing an audience through narrative is effective; for example, Ama-zon's founder and CEO has required executive meetings be switched to a story-telling approach via the "narrative structure" [3]. Jeff Bezos has banned PowerPoints during executive meetings, rather, "... he revealed that the "narrative structure" is more effective than PowerPoint" [3]. These examples of storytelling are even more attentive on social media and are oftentimes told by networks of [initially] organized trolls[1]. The rhetoric of communications through storytelling often mimics manipulative behaviors in order to influence the opinion of the audience by seeking to create an interest with the audience that keeps their attention. Otherwise, what good is storytelling with no one to tell the story to?

Studies have shown that public opinion is going to be more effective than bullets and bombs in the future war. Public opinion has become a tool to achieve a goal, may it be favoritism or turmoil. Information warfare can be used to change and shape public opinion, as it was the case during the Ukraine conflict during which the Russian public was influenced in believing that Russia was defending itself while the West was to blame for the conflict [4]. Influence campaigns can also easily be used to create friction aimed at weakening an adversary as demonstrated by the foreign interference of the U.S. 2016 presidential election [5]. Countering ODG mentality is a must to win the new battle of ideas. Today, transnational terrorist groups know that opinions can be influenced and they are using sophisticated techniques to overcome the time and space limitations of conventional influence campaigns by using digital tactics that take advantage of the speed and reach of the Internet. A study conducted by the Defense Academy of the United Kingdom [6] examines the sharing of the beheading videos of hostages by Al-Qaeda as an instance of strategic communication, defined as: "A systematic series of sustained and coherent activities, conducted across strategic, operational and tactical levels, that enables understanding of target audiences, identifies effective conduits, and develops and promotes ideas and opinions through those conduits to promote and sustain particular types of behaviour" [6].

Taking advantage of storytelling via blogs or social networking sites has quickly moved into strategic narratives and the framework for communications. Studying social media networks to identify false narratives or fake news has become easier because of the new digital information environment. Access to data reveals networks

---

[1] A person who disseminates provocative posts on social media for the troll's amusement or because (s)he was paid to do so.

operating in their true nature, often the networks are extremely large, numbering hundreds of thousands of nodes and ties. Steve Borgatti, a renowned social scientist implies that the importance of an organization in a given network is determined by the institutional affiliations it has within the network [7]. The flow of information amongst the institutional affiliates illuminate areas that are not identified without the use of exploratory social network analysis. *Illuminating these affiliations within networks was conducted through the combination of social network analysis and social cyber forensics. These two approaches allowed us to dissect the network, review the narrative approaches, and study how the authors created disagreement amongst the audience and swayed opinions.*

In this chapter, two different networks of online deviant groups (ODGs) are provided as case studies. The first is the North Atlantic Treaty Organization (NATO) Trident Juncture 2015 exercise and the second is the U.S. Army Europe Operation Dragoon Ride 2015. The focus of this chapter is to identify key actors and clusters within the overall networks and to be able to identify and illuminate these dark networks using social network analysis (SNA) and social cyber forensics (SCF). Malcolm Sparrow, emphasized that intelligence agencies do not have the expertise to conduct SNA, "social network analysis has a lot to offer intelligence agencies in this area through its ability to discover who is *central* within organizations, which individual's removal would most effectively disrupt the network, what role individuals are playing, and which relationships are vital to monitor" [8]. In general, this study provides an academic research background to further develop SNA and SCF based methods and procedures for those seeking the truth. Remember, "The main work of a trial attorney is to make a jury like his client." [9]. Our research demonstrates in both studies that the main actor was the trial attorney and the audience was the jury, easily persuaded to follow, like, retweet, and support the social narrative.

2   Literature Review

For this chapter we have reviewed literature of the work that has been previously conducted in the areas of *bots, cyber forensics*, and *SNA*. Specifically, we explain the background of bot research, data carving, and how cyber forensics is used with SNA. Lastly, a review is conducted on the influence assessment in the blogosphere. One of the aforementioned references in the introduction refers to how modern approaches to developing the narrative are being explored. Corporations that fuse social science into the business place are likely to have a competitive edge in their respective markets. A simple search of Amazon's available job postings as of May 13, 2018, showed multiple social science positions. Some of the preferred qualifications for a position as the Senior Research Psychologist identified "...research background in social or cognitive psychology or affective science, particularly with ties to motivation: deep knowledge of regressions, analysis of variance, multilevel models, structural equation models" [10]. Just this example alone provides a basis for continued research in this field as the job market dictates more and more positions to shape behavior change in the future. Reviewing even literature online in a non-traditional sense such as "job postings" allows us an alternative approach to identify the relevance of this field.

Bots Automated social actors/agents or bots are not a new phenomenon. They have been studied previously in literature in a variety of domains, such as Internet Relay Chat (IRC) [11], online gaming, e.g., World of Warcraft (WoW) [12], and more recently behavioral steering through misinformation dissemination on social media [13]. One of the earliest bots emerged in 1993 in an internet protocol that allows people to communicate with each other by text in real time called Eggdrop. This bot had very simple tasks to welcome new participants and warn them about the actions of other users [11]. Shortly thereafter, the use of bots in IRC became very popular due to the simplicity of the implementation in and their ability to scale IRCs [14]. Both evolved over time and the tasks these bots were assigned became more complicated and sophisticated.

Abokhodair et al. studied the use of social bots regarding the conflict in Syria in 2012 [15]. The study focused on one botnet (i.e., a set of bots working together) that lived for six months before Twitter detected and suspended it [15]. The study analyzed the life and the activities of the botnet. Focus was placed on the content of tweets, i.e., they classified the content of the tweets into 12 categories: news, opinion, spam/phishing, testimonial, conversation, breaking news, mobilization of resistance/support, mobilization for assistance, solicitation of information, information provisioning, pop culture, and other. Through their research, the authors were able to answer the question on how the content of the bot tweeting in Arabic or English differ from the non-bot or legitimate users tweeting in Arabic or English? For example, bots tend to share more news articles, less opinion tweets, no testimonial tweets, and less conversational tweets than any other legitimate Arabic or English Twitter user [16]. They also classified bots based on the content posted, time before the bot gets suspended, and type of activity the bot does (tweet or retweet) into the following categories:

1. Core Bots: These bots are further divided into three sub-categories:
   (a) Generator Bots: bots that tweet a lot but seldom retweet anything.
   (b) Short Lived Bots: bots that retweet a lot but seldom tweet. These bot accounts lasted around 6 weeks before Twitter suspended them.
   (c) Long Lived Bots: bots that retweet a lot but seldom tweet. These bot accounts lasted more than 25 weeks before Twitter suspended them.

2. Peripheral Bots: Twitter accounts that are being lured to participate in the dissemination process. Their task is retweeting one or more tweets generated by the core bots [15].

Research on detecting social bots has increased dramatically. In 2010, Chu et al. [17] proposed a classification system to determine whether tweets on Twitter belong to a human, bot, or cyborg (human account use scripts or tools to post on their behalf, like a hybrid account). Over 500,000 accounts were studied to find the difference between human, bots, and cyborg in tweeting content and behavior. Their classifier is comprised of the following four components: (1) Entropy Component: which is used to detect the regularity and periods of users' tweets, (2) Machine Learning Component: which is used to detect spam tweets, (3) Account Properties Component: which help identify bots by checking external URLs ratio in the tweets or checking the tweeting device (web, mobile, or API) to help detecting bots, and (4) Decision Maker Component: which uses the input of the previous three components to determine the type of the user [17].

Wang et al. [18] reviewed the possibility of human detection, suggesting the crowdsourcing of social bot detection to legions of workers. To test this concept, the authors created an Online Social Turing Test platform. The authors assumed that bot detection is a simple task for humans because humans have a natural ability to evaluate conversational nuances like sarcasm or persuasive language and to observe emerging patterns or anomalies but this is yet unparalleled by machines. Using data from Facebook and Renren—a popular Chinese online social network—the authors tested the efficacy of humans—both expert annotators and workers hired online—at detecting social bot accounts simply from the information on their profiles. The authors observed the detection rate for hired workers drops off over time, although it remains good enough to be used in a majority voting protocol. In their experiment, the same profile was shown to multiple workers and the opinion of the majority was used to determine the final verdict.

The derivative of this literature identifies that bots are present in the current information environment. Sophisticated studies indicate that bots are difficult to monitor even as researchers develop advanced detection methods. Bots are continually growing more advanced demonstrating more human-like behavior [19] which makes them harder to detect, especially if they start to inject "bot" opinions into messaging. Findings from the DARPA "Twitter Bot Detection Challenge" show that bots cannot be solely identified using machine learning only, instead a vast enhancement of analytic tools that combine multiple approaches to help in bot detection is needed [20]. Hence in this chapter, we combine SNA and SCF to help in bot identification, especially in dark networks.

Social Cyber Forensics (SCF) For the last three and half decades digital forensics tools have evolved from simple tools, which were used mainly by law enforcement agencies to import tools for detecting and solving corporate fraud [21]. Cyber forensics tools are not new but they are evolving over time to have more capabilities, more exposure to the audience (investigators or public users), and more types and amount of data that can be obtained using each tool. Cyber forensics tools can be traced back to the early 1980s when these tools were mainly used by government agencies, e.g., the Royal Canadian Mounted Police (RCMP) and the U.S Internal Revenue Service (IRS) and were written in Assembly or C language with limited capabilities and less popularity. With time these tools got more sophisticated and in the mid of 1980s these tools were able to recognize file types as well as retrieve lost or deleted files, e.g., XtreeGold and DiskEdit by Norton. In 1990s these tools became more popular and also have more capabilities, e.g., they can recover deleted files and fragments of deleted files such as Expert Witness and Encase [22]. Nowadays, many tools are available to the public that enable them to collect cyber forensics data and visualize it in an easy to understand way, e.g., Maltego tool (developed by Paterva Ltd. available at www.paterva.com).

Social network forensics tools collect data in many different ways, e.g., crawling by using the social network APIs, extract artifacts from local web browsers cache, or sniffing on unencrypted Wi-Fi's (active attacks), or with ARP spoofing on LANs, or using a third party extension for the social network in combination with a traditional crawler component (friend in the middle attack) [23].

Research by Noora et al. [24] obtains cyber forensics evidence from social media applications that are installed on smartphones. Their research was testing whether the activities conducted through these applications were stored on the device's internal memory or not. They used three major social media apps, i.e., Facebook, Twitter, and MySpace and three devices types, i.e., iPhone, Blackberry, and Android for their experiments. The results show that Blackberry devices do not store any information that can be retrieved by digital forensics tools while iPhone and Android phones store a significant amount of valuable data that can be retrieved [24]. Additional research focused on extracting forensics data of social media from the computer hard disk such as carving artifacts left by the use of Facebook Chat on a computer's hard disk [25].

In this work, we are not creating a tool to collect forensics data from social networks, instead we are using a social cyber forensic analysis tool called Maltego which collects open source information (OSINF) and forensics data. This tool provides a library of transformations for discovery of data from open sources. It helps analyze the real world connections between groups, websites, and affiliations with online services such as Facebook, Flickr, LinkedIn, and Twitter. It also provides the capability to visualize the results in a graph format that is suitable for link analysis.

Social Network Analysis Borgatti implies that the importance of an organization in a given network is determined by the institutional affiliations it has within the network [7]. The flow of information amongst the institutional affiliates will illuminate areas that are not identified without the use of exploratory social network analysis. Often these networks are referred to as *dark networks*. SNA should aid in the overall strategy to identify kinetic and non-kinetic operations, but should not be the definitive component of a stratagem. Applying SNA combined with SCF allows for network illumination of the dark networks.

Common centrality measures such as betweenness, eigenvector, and closeness are used throughout this research. Although these metrics are primary in this research, other areas of SNA are examined as well such as topography, cohesive subgroups, components, and Focal Structure Analysis (FSA).

Networks that portray the shortest paths between the organizations inside of the network, demonstrate betweenness centrality. Further defined, nodes with the closest neighbors are measured by their betweenness centrality [26]. These network measurements are seen in several of the sociograms throughout the chapter. Eigenvector centrality was used to illuminate hierarchy within the organizations. This will display well-connected nodal connections to other well connected nodes [27]. Closeness centrality allows this research to identify the dissemination of information throughout the network. It is imperative to not use closeness as a stand alone metric "This could lead analysts to conclude that certain actors are more important than they really are which of course could lead to using mistaken assumptions when crafting strategies " [28]. It is not unusual that illumination of higher level actors are already common knowledge to the public. Often, bots are used to amplify the messages. Because of the truly hidden ways that bots have been disguised it was necessary to combine SNA and SCF to illuminate the bots.

Focal Structure is an algorithm that was implemented by Şen et al. [29] to discover an influential group of individuals in a large network. FSA is not a community detection algorithm, i.e., in the context of networks, most community detections algorithms try to find nodes that are more densely connected in one part of the network and not that much connected on the other part of the network. These community detection algorithms would suggest that there is a community based on the nodes connection strength (i.e., how closely they are connected). However, FSA is an algorithm that tries to find a key set of nodes that are influential if they are working together (i.e., exist in the network whether they are directly connected or not). These individuals need not to be strongly connected and may not be the most influential actors on their own, but by acting together they form a compelling power.

FSA is a recursive modularity-based algorithm. Modularity is a network structural measure that evaluates the cohesiveness of a network [30]. FSA uses a network-partitioning approach to identify sub-structures or sub-graphs. FSA consists of two parts the first part is a top-down division, where the algorithm identifies the candidate focal structures in the complex network by applying the Louvain method of computing modularity [31]. The second part is a bottom-up agglomeration, where the algorithm stitches the candidate focal structures, i.e., the highly interconnected focal structures, or the focal structures that have the highest similarity values, are stitched together and then the process iterates until the highest similarity of all sibling pairs is less than a given threshold value. Similarity between two structures is measured using Jaccard's Coefficient [29, 32] which results in a value between 0 and 1, where 1 means the two networks are identical, while zero means the two networks are not similar at all. The stitching of the candidate focal structures was done to extract the structures with low densities, i.e., structures contain nodes that are not connected densely [29].

Influence in Blogosphere Blogs provide rich medium for individuals to frame an agenda and develop discourse around it using half-truth or twisting facts to influence the masses. Twitter, however, due to the 280-character limit, is primarily used as a dissemination medium. Bloggers have used Twitter to build an audience (or, followership) and as a vehicle to carry their message to their audience. It is important to understand the disinformation dissemination network on Twitter but it is equally, if not more, important to understand the blog environment and specifically the blogger's influence, engagement with the audience, and motivations for agenda setting.

Identifying influential individuals is a well-studied problem. Many studies have been conducted to identify the influence of a blogger in a community [33–37]. The basic idea of computing the influence a blogger has is to aggregate the influence of their individual blog posts. A blog post having more *in-links* and *comments* indicates that the community is interested in it. *In-links* and *comments* contribute *positively* towards the influence of the posts whereas *out-links* of the blog posts contribute *negatively* towards the influence. Influence can be assessed using a stochastic model with *inbound links*, *comments* and *outbound links* of a post as factors, as proposed in [33]. An alternate approach is to use a modification of *Google page rank* to identify influential posts as well as bloggers [36].

3   Methodology to Study Narratives and Propaganda

In this section, we provide a methodology to study narratives including propaganda that is disseminated on various social media channels during various events. This methodology has been tested on several case studies and provided consistent results. The overall methodology is depicted in Fig. 1. The methodology first starts by domain experts identifying keywords relevant to an event. Second, searching various online social media platforms is conducted to identify an initial seed of data, e.g., Twitter accounts tweeting propaganda about the event, or a YouTube video containing propaganda, or a blog site that contain narratives. Third, using various data collection tools (NodeXL, Scraawl, Web Crawlers (e.g., WebContentExtracor), YouTube APIs, Twitter APIs, TAGs, and Maltego) we extracted the social and communication networks of Twitter users, crawled the blog's data, identified bots, and extracted the metadata associated with the social media accounts of interest. Finally, we conducted a set of analyses on the collected data including:

- Social Cyber Forensics (SCF) analysis to identify relations among various groups, uncover their cross-media affiliation, and identify more groups.
- Social Network Analysis (SNA) to identify leaders of the narrative and identify the role of nodes in the network, e.g., the source of information, brokers, top disseminators, and type of nodes (bot or human account).
- We also conduct various blogs data analyses using our in-house developed Blogtrackers tool (available at: http://blogtrackers.host.ualr.edu) such as sentiment analysis, keywords trends, influential blogs and bloggers, etc.

Fig. 1 The overall research
methodology

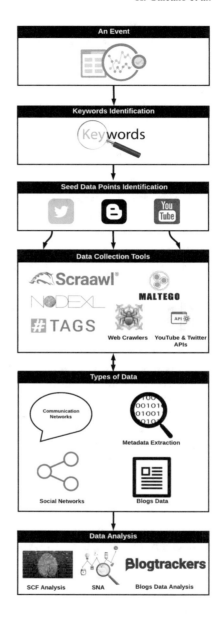

3.1   Case Study 1: Anti-NATO Propaganda During the 2015 Trident Juncture Exercise

What Was the Propaganda On 4 November 2015, the US soldiers along with soldiers from more than thirty partner nations and Allies moved 36,000 personnel across Europe during the 2015 Trident Juncture Exercise (TRJE). The exercise took place in the Netherlands, Belgium, Norway, Germany, Spain, Portugal, Italy, the Mediterranean Sea, the Atlantic Ocean, and also in Canada to prove the capability and readiness of the Alliance on land, air, and maritime. The exercise also demonstrated that the Alliance is equipped with the appropriate capabilities and capacities to face any present or future security issues. In addition to the Partner Nations and Allies, more than Twelve aid agencies, International Organizations, and non-governmental organizations participated in the exercise to demonstrate "NATO's commitment and contribution to a comprehensive approach [38]."

The buildup of the exercise saw a series of competing information maneuvers designed to counter NATO and Allies. Several of these maneuvers are highlighted as examples that were observed in Fig. 2 below. Ranging from narrative hijacking in multiple languages across websites, to community counter NATO meetings, to protests in the streets, and of note an Anti-NATO concert held in Zaragoza. All of which were pushed via social platforms months before the exercise and created an information deficit that NATO had to fill.

Many opponent groups launched campaigns on Twitter, Blogs, Facebook, and other social media platforms that encouraged citizens to protest against the exercise

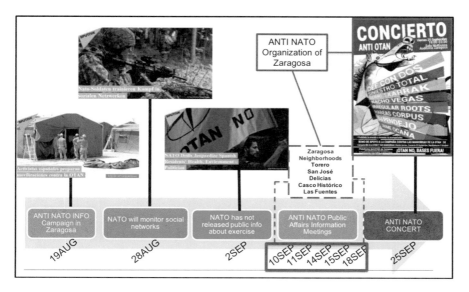

Fig. 2 Examples of Anti NATO activities that were observed in the information environment prior to the exercise [39–41]

or do violent acts. We identified six groups by searching their names on various social media platforms to identify their Twitter and blogging profiles (followed our proposed methodology). These six groups propagated their messages on social media inviting people to act against NATO and TRJE 2015 exercise. Next, we provide a description of the dataset along with our findings.

Data Collection An initial set of twelve blog sites were identified that the groups use to develop narratives against the TRJE 2015 exercise. We were also able to identify Twitter handles used to steer the audience from Twitter to their blogs. We identified an initial set of 9 Twitter accounts used by the six groups. We used Twitter API through a tool called NodeXL to collect a network of replies, mentions, tweets, friends, and followers for all the nine Twitter accounts and whoever is connected to them with any one of the aforementioned relationships for the period 8/3/2014 to 9/12/2015. The dataset file we obtained contains 10,805 friends/followers, 68 replies, 654 tweets, 1365 mentions, 9129 total nodes, and 10,824 total edges. The twitter handles, blogs, and names of the groups studied in this research are publicly available. However, in order to ensure their privacy, we do not disclose them here.

Metadata Extraction We used Maltego which is an open source information gathering and forensics application. Maltego can extract Google Analytics IDs from blog sites. Google Analytics is an online analytics service that allows a website owner to gather statistics about their website visitors such as their browser, operating system, and country among other metadata. Multiple sites can be managed under a single Google analytics account. The account has a unique identifying "UA" number, which is usually embedded in the website's HTML code [42]. Using this identifier other blog sites that are managed under the same UA number can be identified. This method was reported in Bazzell's Open Source Intelligence Techniques: Resources for searching and analyzing online information [43]. Using Maltego we inferred the connections among blog sites and identified new sites that were previously undiscovered.

We used a seed set of 12 blog sites to discover other blogs that are connected to them using Maltego as explained earlier. We used the tool in a snowball manner to discover other blog sites. We were able to identify additional 9 blogs that were connected to the initial seed blogs by the same Google analytics IDs. These newly identified websites have the same content published on different portals and sometimes in different languages. For example, a website written in English may also have another identical version but written in another language that is native to the region. Such blogs are also known as *bridge blogs* [44]. Additional public information such as the IP addresses, website owner name, email address, phone numbers, and locations of all the websites was reviewed. We obtained three clusters of websites based on their geolocation. These clusters are helpful to know the originality of the blog sites, which would help an analyst understand the propaganda that is being pushed by the specific blog site. Cluster 1 contains one website that is located in Russia, Cluster 2 has 8 websites located in USA, and Cluster 3 has 12 blog sites located in Spain, Cayman Islands, UK, and Germany. From the initial 12 blog sites we grew to 21 blog sites, 6 locations, and 15 IP addresses. All the blog sites we identified during this study were crawled and their data is stored in a database that the Blogtrackers tool can access and analyze.

Identifying Influential Information Actors Using SNA In addition to extracting metadata using Maltego to find other related blog sites used by the group to disseminate their propaganda, we applied SNA such as indegree centrality (to assess popular nodes) outdegree centrality (to assess information sources or gregarious nodes), betweenness centrality (to assess information brokers or bridges) to find the most important nodes in the network by activity type we also applied various community detection measures such as modularity (to assess the quality of the clusters), etc. Using NodeXL we were able to find the most used hashtags during the time of the exercise (i.e., the hashtags occurred the most in the collected tweets). This helps in targeting the same audience if counter narratives were necessary to be pushed to the same audience. In addition to that, we found the most tweeted URLs in the graph. This gives an idea about the public opinion concerns. Finally, we found the most used domains, which helps to know where the focus of analysis should be directed, or what other media platforms are used. For example, two of the top 10 hashtags that were used during the TRJE 2015 exercise were #YoConvoco (that translates to "I invite" using Google translation service) and #SinMordazas (that translates to "No Gags"). These two hashtags were referring to a campaign that is asking people for protests and civil resistance or civil disobedience. Also, investigating the top 10 URLs that were shared the most in the dataset reveals that these URLs were links to websites that are mobilizing people to raise objections on using taxpayers' money to fund military spending on wars.

Identifying Powerful Groups of Individuals Affecting Cyber Propaganda Campaign using FSA We divided our network (9129 nodes and 10,824 unique edges) into two type namely, the *social network*, derived from friends and follower's relations and the *communication network*, derived from replies and mentions relations. We ran the FSA algorithm on these two networks to discover the most influential group of nodes.

- Running FSA on the social network resulted in 1 focal structure with 7 nodes. These 7 nodes are in fact among the nine anti-NATO seed nodes we started with and are very tightly knit (i.e., they exert mutually reciprocative relationships). This indicates a strong coordination structure among these 7 nodes, which is critical for conducting information campaigns.
- Running FSA on the communication network resulted in 3 focal structures with a total of 22 nodes. The same 7 accounts (out of the 9 seed accounts) found in the social network focal structures are distributed in these 3 focal structures. This gives those 7 accounts more power/influence than other nodes in the network because they are found in the focal structures of both networks, i.e., the communication and social network. The rest of the nodes (i.e., the additional 15 accounts) found in these 3 focal structures of the communication network are new nodes. These are important because they are either leaders or part of key groups conducting propaganda campaigns.

Analyzing Blogs Data Using Blogtrackers Using SCF analysis and SNA as explained in the previous sections, we were able to identify a total of 21 blog sites of interest. We trained web crawlers to collect data from these blogs and store the data in Blogtrackers database. Then we performed the following analysis:

1. We explore the collected dataset by generating the traffic pattern graph using Blogtrackers. We ran the analysis for the period of August 2014 to December 2015. We observed a relatively higher activity in these blogs from September 2015 to December 2015, the period around the TRJE 2015,
2. We generated a keyword trends graph for the following keywords: 'anti nato', 'trident juncture', 'nato' (as shown in Fig. 3). The keyword trend for the 'anti nato' completely aligned with the traffic pattern graph indicating the posts actually had 'anti nato' keyword in it. We also observed that trend for 'anti nato' was consistently higher than 'nato' for this time period indicating there was more negative sentiment towards NATO in these blogs,
3. We ran the sentiment analysis in Blogtrackers for the same period and observed more negative sentiment than positive sentiment in the blogs,
4. We ran the influential posts analysis in Blogtrackers to identify posts with high influence. In other words, we wanted to identify what resonates with the community most, or which narratives are affecting the people most. The influence score was calculated using of a stochastic model [33] with *inbound links*, *comments* and *outbound links* of a post as factors. The most influential

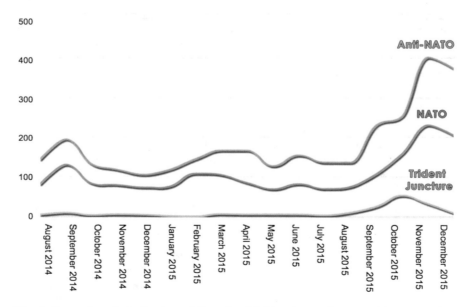

Fig. 3 Keyword trends for "anti nato," "nato," and "trident juncture" generated by Blogtrackers depicting the occurrence of these keywords over the time period

post was an Italian blog post from the 'nobordersard' blog. Upon translation to English we found the post to be highly propaganda-riddled. The blogger used two of the conventional propaganda techniques [45] called "Name Calling" (associating a negative word to damage the reputation) and "Plain Folks" (presenting themselves as ordinary people or general public to gather support for their cause or ideology). The blog post used phrases like: "NATO exercise was contributing to pollution and exploiting resources". It also categorizes this exercise as an act of militarization of territories to train for war. Furthermore, the blog was asking people to protest against the exercise.

3.2 Case Study 2: Anti-NATO Propaganda During the 2015 Dragoon Ride Exercise

What Was the Propaganda On 21 March 2015, US soldiers assigned to the 3rd Squadron, 2nd Cavalry Regiment in Estonia, Latvia, Lithuania, and Poland as part of Operation Atlantic Resolve began Operation Dragoon Ride. US troops, nicknamed 'Dragoons', initiated a military movement which stretched from the Baltics to Germany, crossing five international borders and covering more than 1100 miles. This mission exercised the unit's maintenance and leadership capabilities and also demonstrated the freedom of movement that exists within NATO [46].

Many opponent groups launched campaigns to protest the exercise, e.g., 'Tanks No Thanks' [47], which appeared on Facebook and other social media sites, promising large and numerous demonstrations against the US convoy [48]. Czech President Milos Zeman expressed sympathy with Russia; his statements were echoed in the pro- Russian English language media and the Kremlin financed media, i.e., Sputnik news [49]. The RT website also reported that the Czechs were not happy with the procession of the "U.S.Army hardware" [47]. However, thousands of people from the Czech Republic welcomed the US convoy as it passed through their towns, waving US and NATO flags, while the protesters were not seen.

During that time many bots were disseminating propaganda, asking people to protest and conduct violent acts against the US convoy. A group of these bots was identified using Scraawl (available at www.scraawl.com), an online social media analysis product developed for bot detection and discourse analysis. It's an easy-to-use discovery tool of Intelligent Automation, Inc. for open source information. The link will provide you a free test subscription, if you'd like to try it out yourself. We collected data on this network of bots and studied its structure in an attempt to understand how they operated. Next, we provide a description of the dataset and our findings.

Data Collection We collected data for the period between 8 May 2015 and 3 June 2015 of 90 Twitter accounts that were identified by Scraawl as bots known to disseminate propaganda during the Dragoon Ride Exercise. Out of the 90 Twitter accounts we were able to collect data from 73 accounts. We were not able to collect

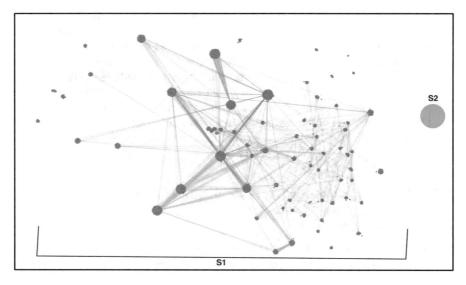

Fig. 4 Two sub-networks, S1 and S2. S1 is un-collapsed while S2 is collapsed. Edges in blue denote mutually reciprocal relations (bidirectional edges) while edges in red color denote non-reciprocal relations (unidirectional edges)

data for 17 Twitter accounts because the accounts had been either *suspended, did not exist*, or were *set to private*. Data was collected using NodeXL—an excel plugin for social media data collection and analysis— that included: friend and follower relations, tweet, mention, and reply relations. This resulted in 24,446 unique nodes and 31,352 unique edges. An 'edge' is a 'relationship', which can be a tweet, retweet, mention, reply, or friendship between two nodes (Twitter accounts). We obtained 50,058 non-unique edges with 35,197 friends and followers edges, 14,428 tweet edges, 358 mention edges, and 75 reply edges.

Analysis of Case Study 2 We analyzed the friend/follower networks (social network) of the bot accounts. We applied the Girvan-Newman clustering algorithm [30] to this network and found that the network had two clusters, S1 and S2, as shown in Fig. 4.

The clusters are the same as the components in this graph. The smaller S2 cluster, containing only a triad of nodes, was rejected from further analysis, as it did not contribute much to the information diffusion. Since the larger S1 cluster contained the majority of nodes, we examined this sub-network further.

Zooming in for a closer examination of the S1 cluster revealed that the members of that network were more akin to a syndicate network, i.e., a network that has dense connections among their members and inter-group connections with the other nodes and do not have a most central node, i.e., no hierarchy. Further examination of the nodes in S1 revealed a mutually reciprocated relationship (the nodes followed each other), suggesting that the principles of '*Follow Me and I Follow You*'

(FMIFY) and 'I Follow You, Follow Me' (IFYFM)—a well-known practice used by Twitter spammers for 'link farming', or quickly gaining followers [50, 51] were in practice—a behavior that was also observed during other study we conducted on the Crimean Water Crisis botnet [52, 53].

This network had no central node or no start-shaped network. In other words, there was no single node feeding information to the other bots, or seeder of information (this was determined using indegree centrality measure). This indicated the absence of a hierarchical organizational structure in the S1 network, in other words no seeder was identified/observed. In cases where the seeder is not easily identifiable, other, more sophisticated methods are warranted to verify if this behaviour truly does not exist. Although there might not be a single most influential node, a group of bots may be coordinating to make an influential group. To study this behaviour further, we applied the Focal Structures Analysis (FSA) approach to find if any influential group of bots existed [54].

FSA has been tested on many real world cases such as the Saudi Arabian Women's Right to Drive campaign on Twitter [55] and the 2014 Ukraine Crisis when President Viktor Yanukovych rejected a deal for greater integration with the European Union and three big events followed—Yanukovych was run out of the country in February, Russia invaded and annexed Crimea in March, and pro-Russian separatist rebels in eastern Ukraine brought the relationship between Russia and the West to its lowest point since the Cold War. Applying focal structures during the two aforementioned examples revealed interesting findings. It was proven that during the Saudi Arabian Women's Right to Drive Twitter campaign on 26 October 2013 the focal structures were more interactive than average individuals in the evolution of a mass protest, i.e., the interaction rate of the focal structures was significantly higher than the average interaction rate of random sets of individuals. It was also proven that focal structures were more interactive than communities in the evolution of a mass protest, i.e., the number of retweets, mentions, and replies increases proportionally with respect to the followers of the individuals in communities [29]. Applying the FSA approach to the Ukraine-Russia conflict also revealed an interesting finding. By applying FSA to a blog-to-blog network, Graham W. Phillips [56]—a 39-year-old British journalist and blogger—was found to be involved in the only focal structure of the entire network along with ITAR-TASS, the Russian News Agency, and Voice of Russia, the Russian government's international radio broadcasting service. Even though other central and well-known news sources, such as the Washington Post and The Guardian, were covering the events, Phillips was actively involved in the crisis as a blogger and maintained a single-author blog with huge influence that compared with some of the active mainstream media blogs. Phillips covered the 2014 Ukraine crisis and became a growing star on Kremlin-owned media. He set out to investigate in a way that made him a cult micro-celebrity during the crisis—by interviewing angry people on the street for 90 s at a time [57].

While the bot detection methodology used in this study is not 100% accurate, we used tools that are commonly used by government agencies. A manual check of accounts identified as bots by the tools used served as an additional verification.

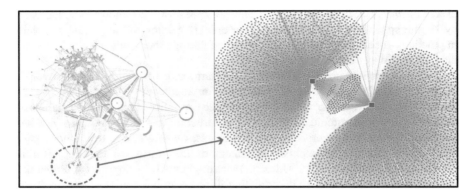

Fig. 5 The social network (friends/followers network) of the botnets. The focal structure analysis approach helped in identifying a highly sophisticated coordinating structure, which is marked inside the blue circle in the figure on left. Upon zooming-in on this structure (displayed on the right), two bots were identified as the seeders in this focal structure. The seeder bots are depicted in blue

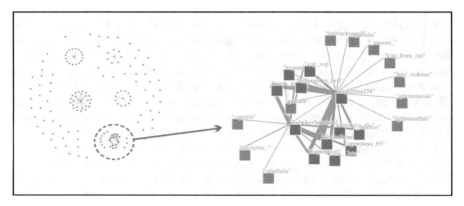

Fig. 6 Communication network (tweets, mentions, and replies network) of the botnets. Ten nodes were communicating the most with the two most influential bots in the network

We ran the FSA approach on the Dragoon Ride data to discover the most influential set of bots or the seeders of information in the S1 community. By applying FSA to the social network of these bots we obtained one focal structure containing two nodes (see Fig. 5). These two nodes form the most influential set of bots in the network, i.e., by working together those two bots had a profound impact on the dissemination of propaganda.

We further applied FSA to the bots' communication network, i.e., tweets, mentions, and replies network to identify who are the most communicative nodes in this network (see Fig. 6). We obtained one focal structure containing 12 nodes. Ten nodes were 'real people nodes', i.e., nodes that communicated the most with bots (potential seeders of information), while the other two nodes were the bots identified as the most influential nodes in the friends and followers network.

In this case study, deviant groups used a sophisticated tool to disseminate their propaganda and speed up the dissemination process by using botnets. These botnets were very sophisticated compared to a study we previously conducted on the use of social bots during the Crimean water crisis in 2014 [52, 53]. The network structure of the botnets in the latter case is much more complex than in the former. Botnets in the Dragoon Ride exercise case required a more sophisticated approach to identify the organizers or seeders of information, i.e., it required applying FSA to both the social network (friends/followers network) and the communication network (tweets, replies, and mentions network). The evolution of complexity in the bots' network structures confirms the need for a systematic study of botnet behavior to develop sophisticated approaches/techniques or tools that can deal with predictive modelling of botnets.

4   Conclusion

In conclusion, the rapid advancement of technology has made people more connected than ever before. Internet, especially social media, has enabled the flow of information at unprecedented rates. This amplification is observed more in the spread of misinformation, fake or inaccurate news, and propaganda. Conducting deviant acts has become more convenient, effective, and rapid. Deviant groups can coordinate cyber campaigns in order to achieve strategic goals, influence mass thinking, and steer behaviors or perspectives about an event in a highly coordinated and sophisticated manner that remains largely undetected.

In this chapter, we provided two important and detailed case studies, namely the NATO's 2015 Trident Juncture Exercise (TRJE 2015) and 2015 Dragoon Ride Exercise. We study the online deviant groups (ODGs) and their behavior in conducting deviant acts, especially disseminating propaganda against NATO during the two exercises. We analyzed situational awareness of the real-world information environment in/around those events by employing computational social network analysis and social cyber forensics informed methodologies. These methodologies help identify information competitors who seek to take the initiative and the strategic message away from the main event in order to further their own agenda. We describe our methodology, analysis (node-level, group-level analysis, and content-level), and results obtained in both case studies. We further study how ODGs use social media in coordinating cyber propaganda campaigns. The research offered many interesting findings and were of great benefit to NATO and U.S. forces participating in both exercises on the ground.

Acknowledgment This research is funded in part by the U.S. National Science Foundation (IIS-1636933, ACI-1429160, and IIS-1110868), U.S. Office of Naval Research (N00014-10-1-0091, N00014-14-1-0489, N00014-15-P-1187, N00014-16-1-2016, N00014-16-1-2412, N00014-17-1-2605, N00014-17-1-2675), U.S. Air Force Research Lab, U.S. Army Research Office (W911NF-16-1-0189), U.S. Defense Advanced Research Projects Agency (W31P4Q-17-C-0059), the Jerry L. Maulden/Entergy Fund at the University of Arkansas at Little Rock, the Arkansas Research Alliance, and Creighton University's College of Arts and Sciences. Any opinions, findings, and conclusions or recommendations expressed in this material are those of the authors and do not necessarily reflect the views of the funding organizations. The researchers gratefully acknowledge the support.

References

1. Professor describes Atari's impact on gaming world. Technique, http://nique.net/life/2009/02/20/professor-describes-ataris-impact-on-gaming-world/. Accessed 1 May 2018
2. M. Zuckerberg, Biography.com, https://www.biography.com/people/mark-zuckerberg-507402. Accessed 3 May 2018
3. C. Gallo, Jeff Bezos banned powerpoint in meetings. His replacement is brilliant. Inc.com (2018), https://www.inc.com/carmine-gallo/jeff-bezos-bans-powerpoint-in-meetings-his-replacement-is-brilliant.html. Accessed 28 Apr 2018
4. D. Volkov, Supporting a war that isn't: Russian public opinion and the Ukraine conflict. Carnegie Moscow Center (2015), https://carnegie.ru/commentary/61236. Accessed 9 Jan 2019
5. D.V. Gioe, Cyber operations and useful fools: the approach of Russian hybrid intelligence. Intell. Natl. Secur. **33**, 954–973 (2018). https://doi.org/10.1080/02684527.2018.1479345
6. S. Tatham, *Strategic Communication: A Primer* (Defence Academy of the United Kingdom, Conflict Studies Research Centre, Camberley, 2008)
7. S.P. Borgatti, Centrality and network flow. Soc. Networks **27**, 55–71 (2005). https://doi.org/10.1016/j.socnet.2004.11.008
8. M.K. Sparrow, The application of network analysis to criminal intelligence: an assessment of the prospects. Soc. Networks **13**, 251–274 (1991). https://doi.org/10.1016/0378-8733(91)90008-h
9. R.B. Cialdini, *Influence: Psychology of Persuasion* (Collins Business, New York, 1993)
10. Senior Research Psychologist – Connections, amazon.jobs, https://www.amazon.jobs/en/jobs/655074/senior-research-psychologist-connections. Accessed 20 May 2018
11. R.A. Rodríguez-Gómez, G. Maciá-Fernández, P. García-Teodoro, Survey and taxonomy of botnet research through life-cycle. ACM Comput. Surv. **45**, 1–33 (2013). https://doi.org/10.1145/2501654.2501659
12. M.S. Ackerman, J. Muramatsu, D.W. Mcdonald, Social regulation in an online game, in *Proceedings of the 16th ACM International Conference on Supporting Group Work - GROUP 10* (2010). https://doi.org/10.1145/1880071.1880101
13. S. Hegelich, D. Janetzko, Are social bots on Twitter political actors? Empirical evidence from a Ukrainian social botnet, in *Proceedings of the Tenth International AAAI Conference on Web and Social Media (International AAAI Conference on Web and Social Media (ICWSM-16)* (2016)
14. A. Karasaridis, B. Rexroad, D. Hoeflin, Wide-scale botnet detection and characterization, in *Proceedings of the First Conference on First Workshop on Hot Topics in Understanding Botnets (HotBots'07)* (USENIX Association, Berkeley, CA, 2007), p. 7-7
15. N. Abokhodair et al., Dissecting a social botnet: growth, content and influence in Twitter, in *Proceedings of the 18th ACM Conference on Computer Supported Cooperative Work & Social Computing, CSCW '15*, ed. by D. Cosley et al. (ACM, New York, 2015), pp. 839–851

16. S. Al-khateeb, N. Agarwal, Examining botnet behaviors for propaganda dissemination: a case study of ISIL's beheading videos-based propaganda, in *Proceedings of the Behavior Analysis, Modeling, and Steering (BEAMS 2015) co-located with the IEEE International Conference on Data Mining (ICDM 2015)* (2015)
17. Z. Chu et al., Who is tweeting on Twitter: human, bot, or cyborg? in *Conference, 2010 Annual Computer Security Applications Conference (ACSAC 2010), Austin, TX, USA - December 06 - 10, 2010* (ACM, New York, 2010), pp. 21–30
18. G. Wang, M. Mohanlal, C. Wilson, X. Wang, M. Metzger, H. Zheng, B.Y. Zhao, Social turing tests: crowdsourcing sybil detection, in *The Network and Distributed System Security Symposium (NDSS)* (The Internet Society, 2013)
19. E. Ferrara, O. Varol, C. Davis, et al., The rise of social bots. Commun. ACM **59**, 96–104 (2016). https://doi.org/10.1145/2818717
20. V. Subrahmanian, A. Azaria, S. Durst, et al., The DARPA Twitter bot challenge. Computer **49**, 38–46 (2016). https://doi.org/10.1109/mc.2016.183
21. N. Alherbawi, Z. Shukur, R. Sulaiman, Systematic literature review on data carving in digital forensic. Procedia Technol. **11**, 86–92 (2013). https://doi.org/10.1016/j.protcy.2013.12.165
22. K. Oyeusi, *Computer Forensics* (London Metropolitan University, London, 2009)
23. M. Mulazzani, M. Huber, E. Weippl, Social network forensics: tapping the data pool of social networks, in *Eighth Annual IFIP WG*, vol. 11 (2012)
24. N.A. Mutawa, I. Baggili, A. Marrington, Forensic analysis of social networking applications on mobile devices. Digit. Investig. **9**, S24 (2012). https://doi.org/10.1016/j.diin.2012.05.007
25. N.A. Mutawa, I.A. Awadhi, I.M. Baggili, A. Marrington, Forensic artifacts of Facebook's instant messaging service, in *2011 International Conference for Internet Technology and Secured Transactions* (2011), pp. 771–776
26. K.M. Carley, J. Reminga, ORA: organization risk analyzer. Technical Report, Carnegie Mellon University, School of Computer Science, Institute for Software Research International (2004), http://www.casos.cs.cmu.edu/publications/papers/carley_2004_oraorganizationrisk.pdf. Accessed 21 May 2018
27. G. Cheliotis, Social network analysis. LinkedIn SlideShare (2010), https://www.slideshare.net/gcheliotis/social-network-analysis-3273045. Accessed 25 May 2018
28. S.F. Everton, Strategic options for disrupting dark networks, in *Disrupting Dark Networks*, (Cambridge University Press, New York, 2012), pp. 32–46. https://doi.org/10.1017/cbo9781139136877.004
29. F. Şen, R. Wigand, N. Agarwal, et al., Focal structures analysis: identifying influential sets of individuals in a social network. Soc. Netw. Anal. Min. **6**, 17 (2016). https://doi.org/10.1007/s13278-016-0319-z
30. M. Girvan, M.E.J. Newman, Community structure in social and biological networks. PNAS **99**(12), 7821–7826 (2002)
31. V.D. Blondel, J.-L. Guillaume, R. Lambiotte, E. Lefebvre, Fast unfolding of communities in large networks. J. Stat. Mech: Theory Exp. **2008**, P10008 (2008). https://doi.org/10.1088/1742-5468/2008/10/p10008
32. P. Jaccard, The distribution of the flora in the alpine zone. New Phytol. **11**(2), 37–50 (1912)
33. N. Agarwal, H. Liu, L. Tang, P.S. Yu, Identifying the influential bloggers in a community, in *Proceedings of the International Conference on Web Search and Web Data Mining - WSDM 08* (2008). https://doi.org/10.1145/1341531.1341559
34. N. Agarwal et al., Modeling blogger influence in a community. Soc. Netw. Anal. Min. **2**(2), 139–162 (2012)
35. S. Kumar et al., Convergence of influential bloggers for topic discovery in the blogosphere, in *International Conference on Social Computing, Behavioral Modeling, and Prediction*, ed. by S.-K. Chai et al. (Springer, Berlin 2010), pp. 406–412
36. A. Java, P. Kolari, T. Finin, T. Oates, Modeling the spread of influence on the blogosphere, in *Proceedings of the 15th International World Wide Web Conference* (2006), pp. 22–26

37. K.E. Gill, How can we measure the influence of the blogosphere, in *WWW 2004 Workshop on the Weblogging Ecosystem: Aggregation, Analysis and Dynamics* (2004)
38. A.J. Girao, Tried and tested, in *NATO Summit 2016 – Strengthening Peace and Security* (2016), pp. 105–107
39. Sputnik, Activistas españoles preparan movilizaciones contra la OTAN. Sputnik Mundo (2015), http://mundo.sputniknews.com/espana/20150819/1040501131.html. Accessed 20 Aug 2015
40. Sputnik, Nato-Soldaten trainieren Kampf in sozialen Netzwerken. Sputnik Deutschland (2015), http://de.sputniknews.com/militar/20150828/304057978.html. Accessed 2 Sep 2015
41. Sputnik, NATO drills jeopardize Spanish residents' health, environment - politician. Sputnik International (2015), http://sputniknews.com/military/20150902/1026475239/nato-drills-zaragoza.html. Accessed 2 Sep 2015
42. L. Alexander, Open-source information reveals pro-kremlin web campaign. Global Voices (2015), https://globalvoices.org/2015/07/13/open-source-information-reveals-pro-kremlin-web-campaign. Accessed 21 May 2018
43. M. Bazzell, *Open Source Intelligence Techniques: Resources for Searching and Analyzing Online Information* (Createspace Independent Publishing, Charleston, 2016)
44. B. Etling, J. Kelly, R. Faris, J. Palfrey, *Mapping the Arabic blogosphere: politics, culture, and dissent*, vol 6 (Berkman Center for Internet & Society, Cambridge, 2009)
45. R.B. Standler, Propaganda and how to recognize it (RBS0), www.rbs0.com/propaganda.pdf. Accessed 2 Sep 2005
46. Operation Atlantic Resolve Exercises Begin in Eastern Europe. U.S. Department of Defense. https://www.defense.gov/News/Article/Article/604341/operation-atlantic-resolve-exercises-begin-in-eastern-europe/. Accessed 21 May 2018
47. 'Tanks? No thanks!': Czechs unhappy about US military convoy crossing country. RT International. https://www.rt.com/news/243073-czech-protest-us-tanks/. Accessed 21 May 2018
48. D. Sindelar, U.S. Convoy, in Czech Republic, Real-Life Supporters Outnumber Virtual Opponents, Radio Free Europe/Radio Liberty (2015)
49. Sputnik, Czechs plan multiple protests of US Army's 'operation dragoon ride'. Sputnik International (2015), https://sputniknews.com/europe/201503281020135278/. Accessed 21 May 2018
50. S. Ghosh, B. Viswanath, F. Kooti, et al., Understanding and combating link farming in the twitter social network, in *Proceedings of the 21st international conference on World Wide Web - WWW 12* (2012). https://doi.org/10.1145/2187836.2187846
51. V. Labatut, N. Dugue, A. Perez, Identifying the community roles of social capitalists in the Twitter network, in *2014 IEEE/ACM International Conference on Advances in Social Networks Analysis and Mining (ASONAM 2014)* (2014). https://doi.org/10.1109/asonam.2014.6921612
52. S. Al-Khateeb, N. Agarwal, Understanding strategic information manoeuvres in network media to advance cyber operations: a case study analysing pro-Russian separatists' cyber information operations in Crimean water crisis. J. Balt. Secur. **2**, 6 (2016). https://doi.org/10.1515/jobs-2016-0028
53. N. Agarwal, S. Al-Khateeb, R. Galeano, R. Goolsby, Examining the use of botnets and their evolution in propaganda dissemination. Def. Strateg. Commun. **2**, 87–112 (2017). https://doi.org/10.30966/2018.riga.2.4
54. Şen F, Wigand R, Agarwal N, et al., Focal Structure Analysis in Large Biological Networks, in *IPCBEE (2014 3rd International Conference on Environment Energy and Biotechnology)*, vol. 70 (IACSIT Press, Singapore, 2014), p. 1
55. S. Yuce et al., Studying the evolution of online collective action: Saudi Arabian Women's "Oct26Driving" Twitter campaign, in *International Conference on Social Computing, Behavioral-Cultural Modeling, and Prediction*, ed. by W.G. Kennedy et al. (Springer, Cham, 2014), pp. 413–420

56. Graham Phillips is a British national contracted as a stringer by the Russian Times (RT). He has produced numerous videos, blogs, and stories in and around eastern Ukraine. He speaks and writes in Russian and English in his reports. He recently spent time covering the World Cup in Brazil for RT and has re-entered Eastern Ukraine as of July 2014. RT reported on that Phillips was deported from Ukraine because he works for RT. He will not be allowed to re-enter Ukraine for three years.
57. M. Seddon, How a British blogger became an unlikely star of the Ukraine conflict - and Russia Today. BuzzFeed, https://www.buzzfeed.com/maxseddon/how-a-british-blogger-became-an-unlikely-star-of-the-ukraine?utm_term=.poaNaBd7w#.ikzWz90Ny. Accessed 21 May 2018

You Are Known by Your Friends: Leveraging Network Metrics for Bot Detection in Twitter

David M. Beskow and Kathleen M. Carley

Abstract Automated social media *bots* have existed almost as long as the social media platforms they inhabit. Although efforts have long existed to detect and characterize these autonomous agents, these efforts have redoubled in the recent months following sophisticated deployment of bots by state and non-state actors. This research will study the differences between human and bot social communication networks by conducting an account snow ball data collection, and then evaluate network, content, temporal, and user features derived from this communication network in several bot detection machine learning models. We will compare this model to the other models of the *bot-hunter* toolbox as well as current state of the art models. In the evaluation, we will also explore and evaluate relevant training data. Finally, we will demonstrate the application of the *bot-hunter* suite of tools in Twitter data collected around the Swedish National elections in 2018.

1 Introduction

Automated and semi-automated social media accounts have been thrust into the forefront of daily news as they became associated with several publicized national and international events. These automated accounts, often simply called *bots* (though at times called *sybils*), have become agents within the increasingly global marketplace of beliefs and ideas. While their communication is often less sophisticated and nuanced than human dialogue, their advantage is the ability to conduct timely informational transactions effortlessly at the speed of algorithms. This advantage has led to a variety of creative automated agents deployed for beneficial as well as harmful effects. While their purpose, characteristics, and "puppet masters" vary widely, they are undeniably present and active. Their effect, while difficult if not impossible to measure, is tangible.

D. M. Beskow (✉) · K. M. Carley
School of Computer Science, Carnegie Mellon University, Pittsburgh, PA, USA
e-mail: dbeskow@andrew.cmu.edu; kathleen.carley@cs.cmu.edu

© Springer Nature Switzerland AG 2020   53
M. A. Tayebi et al. (eds.), *Open Source Intelligence and Cyber Crime*, Lecture Notes in Social Networks, https://doi.org/10.1007/978-3-030-41251-7_3

Automated and semi-automated accounts are used for a wide variety of reasons, creating effects that can be positive, nuisance, or malicious. Examples of positive bots include personal assistants and natural disaster notifications. Nuisance bots are typically involved in some type of 'spam' distribution or propagation. The spam content ranges from commercial advertising to the distribution of adult content. Malicious bots are involved in propaganda [52], suppression of dissent [64], and network infiltration/manipulation [8].

Malicious bots have recently gained wide-spread notoriety due to their use in several major international events, including the British Referendum known as "Brexit" [43], the American 2016 Presidential Elections [13], the aftermath of the 2017 Charlottesville protests [35], the German Presidential Elections [55], the conflict in Yemen [7], and recently in the Malaysian presidential elections [4]. These accounts attempt to propagate political and ideological messaging, and at times accomplish this through devious cyber maneuver.

As these bots are used as one line of effort in a larger operation to manipulate the marketplace of information, beliefs, and ideas, their detection and neutralization become one facet of what is becoming known as *social cyber security*. Carley et al. is the first to use this term, and defines it as:

> Social Cyber-security is an emerging scientific area focused on the science to characterize, understand, and forecast cyber-mediated changes in human behavior, social, cultural and political outcomes, and to build the cyber-infrastructure needed for society to persist in its essential character in a cyber-mediated information environment under changing conditions, actual or imminent social cyber-threats. [18]

Within *social cyber security*, bot detection and neutralization are quickly becoming a *cat and mouse* cycle where detection algorithms continuously evolve trying to keep up with ever-evolving *bots*. Early detection algorithms exploited the automated timing, artificial network structure, and unoriginal meta-data of automated accounts in order to identify them. These features are relatively easy for bot puppet-masters to manipulate, and we are now seeing automated accounts that have meaningful screen names, richer profile meta-data, and more reasonable content timing and network characteristics.

We are also seeing an increasing number of accounts that we call "bot assisted" or "hybrid" accounts (also at times called "cyborg" accounts). Although researchers often attempt a binary classification of *bot* or *human*, the reality is that there is a spectrum of automated involvement with an account. Many accounts are no longer strictly automated (all content and social transactions executed by a computer). These accounts will have human intervention to contribute nuanced messaging to two-way dialogue, but will have a computer executing a variety of tasks in the background. Grimme et al. [38] discusses this spectrum in detail, describing how 'social bots' are created, used, and how 'hybridization' can be used to bypass detection algorithms (in their case successfully bypassing the 'Botornot' algorithm discussed later in this paper).

We hypothesize that bots are not involved in social networks and social communication in the same way that humans are, and that this difference is measurable. Like other complex systems (natural ecosystems, weather systems, etc), social interaction and relationships are the result of myriads of events and stimuli in both the real and virtual worlds. Because bots lack real world engagement and social environments, they embed in different networks than humans.

Many bots are programmed to interact with each other as a bot network, and attempt to interact with humans, but many features of these interactions will be 'robotic'. Even 'hybrid' accounts will have some level of artificial and inorganic structure and substance in their communications. This area of bot detection in Twitter is largely unexplored, primarily because the rich network data (both the friends/followers network as well as their conversational network) are very time consuming to collect. We therefore set out to collect the data to characterize the social network(s) and social conversation(s) that a twitter account participates in, describe these networks with various network metrics, leverage these rich network metrics in traditional machine learning models, and evaluate whether the time involved creates substantial value.

1.1 Research Questions

1. Do bot Twitter accounts have fundamentally different conversational network structures than human managed accounts?
2. Do the conversations that surround bot accounts diverge from human conversations in general substance and timing?
3. Can the measured differences between bot and human conversation networks lead to increased accuracy in bot detection?

This paper will begin by discussing past bot detection techniques, as well as summarize historical techniques for extracting features from network structures. Next we discuss our data collection, data annotation, and methodology for creating ego-network metrics. We describe training and testing our bot-hunter machine learning algorithms and present our results. We construct an evaluation to compare all bot-hunter models against the state of the art. Finally, we will demonstrate the application of the bot-hunter suite of tools in the 2018 Swedish National elections, providing a possible workflow to open source intelligence practitioners.

This chapter is an extension of [10], with a focus of extending the feature space beyond network metrics to include content and temporal metrics of the larger ego network. Several of these features are novel, including a cascaded classifier that identifies portion of alters that are likely bots, portion of alters that don't have normal daily rhythms, as well as portion of ego network that produces tweets that are more popular than the account itself. All of these have been documented as attributes of bots, and we've coded them into features in this algorithm. Additionally, we used the larger models to explore several new bot data sets. Finally, this extension will compare all of the bot-hunter suite of tools against state of the art models.

2   Related Work

2.1   Understanding Data Tiers

In earlier research our team proposed a tiered approach to bot detection [11] that mirrors the data tiers introduced below. This tiered approach creates a flexible bot-detection "tool-box" with models designed for several scenarios and data granularities. Tier 0 builds models on a single entity (usually a tweet text or *user* screen name). Tier 1 builds models based on features extracted from the basic Tweet object (and associated *user* object). Tier 2 extracts features from a users' timeline, and Tier 3 (explained in this paper) builds features from the conversation surrounding a user. Higher tier models are generally more accurate, but consume more data and are therefore computationally expensive. Some research requires bot detection at such a scale, that models based on Tier 0 or Tier 1 are the only feasible option. At other times, highly accurate classification of a few accounts is required. In these cases, models based on Tier 2 or Tier 3 data are preferred. This paper proposes an approach to Tier 2–3 bot detection that builds on the previous Tier 0 [12] and Tier 1 [11] research and relies heavily on network metrics collected through single seed snowball sampling. We will view past research in bot detection through the lens of these Tiers (Table 1).

Since the early efforts to conduct bot/spam detection, numerous teams have developed a variety of models to detect these. While similar, these models will differ based on the underlying data they were built on (for example many community detection and clickstream models were developed for Facebook, while the overwhelming majority of models built on Twitter data use Supervised and Unsupervised Machine learning [1]). Even in Twitter bot detection, these models can be grouped by either the models/methods or by the data that they use. We have provided Table 2 to outline the connection between past models and the data that they use.

Adewole et al. [1] reviewed 65 bot detection articles (articles from 2006–2016) and found that 68% involved machine learning, 28% involved graph techniques (note that these include some machine learning algorithms that rely heavily on

Table 1 Four *tiers* of Twitter data collection to support account classification (originally presented in [11])

Tier	Description	Focus	Collection time per 250 accounts	# of Data entities (i.e. tweets)
Tier 0	Tweet text only	Semantics	N/A[a]	1
Tier 1	Account + 1 Tweet	Account meta-data	∼1.9 s	2
Tier 2	Account + Timeline	Temporal patterns	∼3.7 min	200+
Tier 3	Account + Timeline + Friends Timeline	Network patterns	∼20 h	50,000+

[a]This tier of data collection was presented by Kudugunta and Ferrara [47] and assumes the status text is acquired outside of the Twitter API

Table 2 Table of Twitter Bot detection models and the data that they use

Data	Community detection	Machine learning		Crowd sourcing
		Supervised	Unsupervised	
Tier 0 Text		[12, 47]	[48]	
Tier 1 + Profile		[22, 49]	[33]	
Tier 2 + History		[63]	[20]	
Tier 3 + Snowball	[8]	No known research		[66]
Stream		[3, 14]		

network metrics), and 4% involved crowd-sourcing. Below we will summarize the salient works under each of these modeling techniques.

2.2 Machine Learning Techniques

As noted above, Twitter bot detection has primarily used Machine Learning models. The *supervised* machine learning models used for bot detection include Naïve Bayes [22], Meta-based [49], SVM [48], and Neural Network [47]. The *unsupervised* machine learning models used include hierarchical [48], partitional [33], PCA-based [65], Stream-based [53], and correlated pairwise similarity [20]. Most of these efforts leverage data collected from the basic tweet object or user object (what we would define as a *Tier 0* or *Tier 1* model).

In 2014, Indiana University launched one of the more prominent supervised machine learning efforts with the *Bot or Not* online API service [25] (the service was recently rebranded to *Botometer*). This API uses 1150 features with a random forest model trained on a collage of labeled data sets to evaluate whether or not an account is a bot. *Botometer* leverages network, user, friend, temporal, content, and sentiment features with Random Forest classification [28].

In 2015 the Defense Advanced Research Projects Agency (DARPA) sponsored a Twitter bot detection competition that was titled "The Twitter Bot Challenge" [59]. This 4 week competition pitted four teams against each other as they sought to identify automated accounts that had infiltrated the informal Anti-Vaccine network on Twitter. Most teams in the competition tried to use previously collected data (mostly collected and tagged with *honey pots*) to train detection algorithms, and then leverage tweet semantics (sentiment, topic analysis, punctuation analysis, URL analysis), temporal features, profile features, and some network features to create a feature space for classification. All teams used various techniques to identify initial bots, and then used traditional classification models (SVM and others) to find the rest of the bots in the data set.

2.3   Other Techniques

Several other novel bot detection methods exist outside of machine learning and
network based approaches. Wang et al. [66] investigated the idea of Crowd Sourcing
bot detection. While showing limited success, it was costly at scale, and usually
required multiple workers to examine the same account. Another unique type of
unsupervised learning involves algorithms that find and label correlated accounts.
Most bots are not deployed by themselves. Even if not deployed as a united bot-net,
many *bot herders* often task multiple bots to perform the same operations. Chavoshi
et al. [20] has leveraged the semantic and temporal similarity of accounts to identify
bots in an unsupervised fashion, creating the *Debot* model which we will compare
against in our results section.

2.4   Network Based Techniques

Networks are an extremely important part of bots, bot behavior and bot detection.
Aiello et al. [2] discusses the impact of bots on influence, popularity, and network
dynamics. Adewole et al. [1] highlights that network features are robust to criminal
manipulation.

One approach to leveraging network structure involves community based
bot/sybil detection. While community detection has been effectively implemented
on Facebook [67] and Seino Weibo [51], it has only recently been used on Twitter
Data due the strict friend/follower rate limiting discussed above. Only recently
has Benigni et al. [9] used dense subgraph detection to find extremists and their
supporting bots in Twitter.

Most research that uses networks for bot detection with Twitter Data are in fact
creating network based metrics and introducing these features in traditional machine
learning models. As discussed below, the most challenging part of this type of
research is focused on how to build networks from limited data. The closest works
to ours were performed by Bhat and Abulaish [14] in 2013 and [3] in 2016. Both
research efforts used network features along with profile and temporal features from
a Twitter Sample Stream without any snowball sampling enrichment. They created
an egocentric network that involved ego, alters, with links between alters for both
following and mention ego centric networks. Having done this, they calculated content, profile, and social interaction features. Their network features were restricted
to centrality measures, density measures, and weak and strongly connected components. A similar earlier work by Bhat and Abulaish [14] attempts to use community
features (number of communities, core/periphery, foreign in/out degree, etc). This
was applied to both Facebook data and the Enron email data (not to Twitter).

Additionally, the *Botometer* algorithm leverages some network features extracted
from the user timeline. This includes metrics on the *retweet* network, *mention*
network, and *hashtag* co-occurrence network. The metrics include density, degree

distributions, clustering coefficient, and basic network characteristics. The *Botometer* algorithm does not conduct a snowball collection of *friends* or *followers*, but does appear to collect user objects for accounts found in the timeline as a *retweet* or *mention* [28].

2.5 Building Networks with Twitter Data

As noted above, however, it is difficult to quickly build comprehensive network structure with Twitter data due the Twitter API rate limits, primarily associated with collecting friend/follower ties. Researchers have generally used one of two methods to build limited networks.

The first method is used if the research team has a large sample or stream. These samples may be random (collected from the 1% Twitter Sample) or they may be associated with an event or theme (i.e. collecting all Tweets that have a given hashtag like #hurricanesandy). These researcher then build ego-centric networks from this stream, without collecting any additional data from the Twitter API. This has the advantage of speed, and doesn't suffer from issues getting data for suspended accounts. This method, however, will only model a small portion of an account's activity and network. A 1% sample will arguably contain marginal activity for given account, and even topical streams will only contain a small part of an account's activity, given that they are involved in multiple topics and discussions. These small samples may not be rich enough to serve as strong features for machine learning.

The second method that researchers use is to only collect the users *timeline* (history of tweets, up to last 3200). They then build an ego-centric network from this data (variously using replies, retweets, hashtags, urls, and mentions to build networks). This is much richer than the first method, contain all of the users activity, but still lacks any information beyond that individual, providing the limited star graph illustrated in Fig. 1. It doesn't contain the larger conversation(s) that they are participating in. Additionally, a bots's *timeline* is completely managed by the bot *puppet master*, and therefore can be manipulated to avoid detection.

To date our team has not found supervised learning bot detection research that leverages extensive snowball sampling to build ego networks.

2.6 Extracting Features from Social Networks

Evaluating network centrality measures, started by Bavelas in 1948 [6] and effectively clarified by Freeman in 1978 [31], has long been an important metric for evaluating both nodes and networks. According to Freeman, network-level centrality metrics measure the "compactness" of the network. Our model includes several network centrality measures: degree centrality, k-betweenness [5], and eigenvector centrality [45] are used to measure differing "compactness" between human and bot conversation networks.

Several seminal works describe the importance of triadic relationships in social networks [19, 40] and as a foundation for measuring network clustering and groups [42]. The fact that the study of triadic relationship has almost exclusively been contained within the study of social interaction provides evidence that these observed triadic relationships are unique to human behavior. We have therefore included several features based on these triadic relationships, including the full triadic census [41], number of Simmelian Ties [26, 46], and clustering coefficient. We also included reciprocity based on Mislove et al.'s [54] examination of reciprocity in online social networks.

In addition to finding network centrality and triadic structures, network community detection has been an important aspect of network characterization, and is still an active research area. Current group detection techniques generally fall into traditional methods, divisive methods, modularity based methods, statistical inference methods, and dynamic methods [29]. Our community detection features leverage Louvain Clustering [16], which is based on modularity optimization.

Our approach uses network sampling in order to restrict the time of computation. While research in network sampling started in the 1970s with work from [30] and others, the emergence of Online Social Networks (OSN's) increased the size of networks and the need for sampling. Our approach to sampling ego networks was informed by Gjoka [34]. Our sampling uses breadth-first-search (BFS) on the target node. The known bias of BFS is eliminated because we are only conducting two hops from the target (only includes friends of alters).

Finally, the study of ego networks is a special branch of social network analysis that is relevant to our study. In 1972, [37] presented the classic concept of the "Strength of Weak Ties" in ego networks, which [17] clarified is more due to the structural location of ties, and can be measured by effective size, efficiency, and constraint. This informed our use of ego network effective size in our features space. Additionally, Centrality of ego-networks was explored by Freeman [32] in 1982 informing our use of betweenness in the feature space.

2.7 Contributions of This Work

While we discuss above several other research attempts to use network metrics in a bot detection feature space, these have largely relied on the mention network extracted from any Twitter query/stream. Ego-centric networks built on a single stream/query arguably contain only a small subset of the overall account ego network. Researchers have not attempted to build this ego network based on snowball sampling [36] with a seed node since this requires significant time given the extent of the data and the strict API rate limits that Twitter imposes on friend/follower data. Our research has taken the time to build this rich conversational network in a novel way, and then evaluate whether the time and effort render sufficient value.

Having built this extensive network for every account in question, this work attempts to fully exploit all available features, going above and beyond just structural features. These additional features include content, temporal, and user summary features. Adding the full range of additional features allows us to fully evaluate the increased accuracy against the additional computational cost.

This work additionally creates and explores bot detection metrics that require greater effort and sophistication to circumvent. Currently, bot-herders can circumvent current algorithms by changing their screen name, adding account meta-data, spending additional time selecting a unique profile picture, and creating a more realistic tweet inter-arrival time. They can also deploy bots in bot networks, therefore artificially manipulating friend/follower values to appear like they are popular. However, it will arguably require significantly more sophistication to change the centrality, components, or triadic relationships in the conversations that they participate in. By increasing the cost to deploy and operate bots, it may economically force "bot-herders" out of their devious market.

Finally, the *bot-hunter* framework builds on the multi-tiered bot detection approach that we introduced in [11]. This multi-tiered approach provides researchers and government or non-governmental agencies with a "tool-box" of models designed for different classes of bots as well as different scales of data (designed for either high volume of high accuracy). This multi-tiered approach acknowledges that there is not a one-size fits all model/approach that will work for all bot detection requirements. By merging and expanding on past bot detection research, we can create an easy to use "tool box" that can address several bot-detection requirements. The evaluation provided later in this paper will demonstrate that key models in the bot-hunter suite of tools are equivalent or better than state of the art models.

3 Data

Our team used the Twitter REST and Streaming API's to access the data used in this research effort. Details of this process are provided below.

3.1 Overview of Available Data

Research is loosely divided between account-focused data collection strategies and topical or stream based collection strategies. Account based approaches will only use data objects directly tied to the user (user JSON object, user time-line object, etc). Stream-based approaches extract features from a given topical stream or twitter stream sample. These stream based features are often network features, but represent a small fraction of the ego-centered network of a given account. Our research therefore pursues an account based approach to build a fuller representation of the account's ego network.

Researchers must find a balance between speed and richness of data. Past account focused research generally falls into four *tiers*. Table 1 provides a description for each *tier* of data collection, the estimated time it would require to collect this data for 250 accounts, and the amount of data that would be available for feature engineering per account.

3.2   Data Required for Account Conversation Networks

Detailed ego network modeling of a Twitter account's social interactions requires Tier 3 data collection, but to date our team has not found any research that has conducted that level of data collection to model the network structures and social conversations that an automated Twitter account interacts with. In fact, few teams go beyond basic in-degree (follower count) and out-degree (friend count) network metrics found in Tier 1 meta-data. The closest effort to date is the *Botometer* model, which arguably operates at *Tier 2*. By adding the user timeline, *Tier 2* provides limited network dynamics, to include being able to model hashtag and URL co-mentions in a meta-network (see Fig. 1). The resulting *timeline* based network, however, lacks comprehensive links between alters. While the time-line can provide rich temporal patterns, we found that it lacked sufficient structure to model the ego network of an actor.

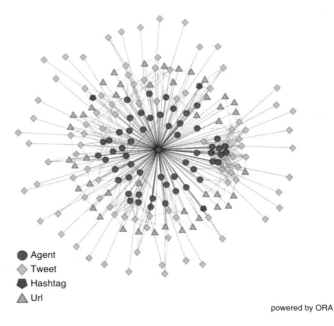

powered by ORA

Fig. 1 Leveraging only user *timeline* provides limited network features in a *star graph*

We set about to build the social network and social conversations that a twitter account is interacting with. We also tried to do this in a way that would expedite the time it takes to collect the data and measure network metrics. Our initial goal was to collect data, build the feature space, and classify an account within 5 min. We selected the 5 min limit in an attempt to process ~250 accounts per day with a single thread

To collect the necessary data, we executed the following steps sequentially:

1. Collect user data object
2. Collect user timeline (last 200 tweets)
3. Collect user followers (if more than 250, return random sample of 250 followers)
4. Collect follower timelines (last 200 tweets)

When complete, this data collection process (illustrated in Fig. 2) creates up to 50,000 events (tweets) that represent the conversation and virtual social interaction that the user and their followers participate in.

The resulting network, while partially built on social network structure (the initial *following* relationship), is primarily focused on the larger conversation they participate in. We initiated the single seed snowball by querying *followers* rather

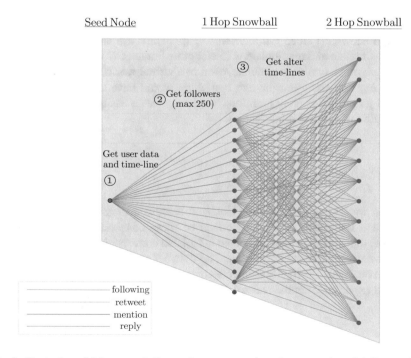

Fig. 2 Illustration of 2-hop snowball sampling: conversation of target node and followers. First get followers of target node (if more than 250, sample followers). Then get timelines of alters. Use timelines to draw connections to accounts that alters *retweet*, *reply*, and *mention*

than *friends* since *followers* are much less controlled by the bot-herder, and contain fewer news and celebrity accounts. We conducted a *timeline* rather than *followers* search for the 2nd hop of the snowball to overcome rate-limiting constraints and to model the conversation network rather than directly model the social network. This single seed snowball process conducts a limited breadth-first-search starting with a single seed and terminating at a depth of 2.

Artificially constraining the max number of alters at 250 was a modeling compromise that facilitates the self-imposed 5 min collect/model time horizon. The choice of 250 allows our process to stay under 5 min, and also represents the upper bound of Dunbar's number (the number of individuals that one person could follow based on extrapolations of neocortex size) [27]. Additionally, in evaluating a sample of 22 million twitter accounts, we found that 46.6% had less than 250 followers. This means that approximately 50% of accounts will have their entire ego network modeled. Bots tend to have fewer followers than human accounts and from the 297,061 annotated bot accounts that we had available for this research, 72.5% of them had fewer than 250 followers. Given that this compromise will only affect 25% of the bot accounts and 50% of all accounts, we felt that it was appropriate.

We used this data to create an agent to agent network where links represent one of the following relationships: mention, reply, retweet. These collectively represent the paths of information and dialogue in the twitter "conversation". We intentionally did not add the *follow/friend* relationships in the network (collected in the first hop of the snowball) since *follow/friend* relationships are an easy metric for bot herders to simulate and manipulate with elaborate bot nets. Complex conversations, however, are much harder to simulate, even in a virtual world. Additionally, adding the *following* links between the ego and alters would have created a single large connected graph. By leaving them out, we were able to easily identify the natural fragmentation of the social interaction.

3.3   Visualizing Conversations

During our initial exploration, we visualized these *conversations* for both human accounts and bot accounts. A comparison of these conversations is provided in Fig. 3. Note that bots tend to get involved in isolated conversations, and the followers of the bot are very loosely connected. The network created from a human virtual interaction on Twitter, is highly connected due to shared friendship, shared interests, and shared experiences in the real world.

3.4   Annotated Data

For annotated bot data, we combined several legacy annotated bot data sets as well as some that our team has annotated during the development of the *bot-hunter*

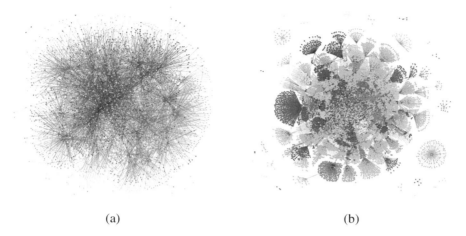

(a)   (b)

Fig. 3 Differences between a human Twitter conversation(s) and a bot Twitter interactions (networks colored by Louvain group) [10]. (**a**) Human *conversation*. (**b**) Bot *conversation*

toolbox. Note that Tier 3 model requires additional collection of friends, followers, and followers timeline, and therefore requires accounts that are not suspended. Several rich annotated bot data sets were used for our Tier 1 and Tier 2 models have a high number of suspended accounts, and therefore were not used for the development of a Tier 3 model. These datasets will still be discussed in the results and evaluation sections since they were used in the development of Tier 1 and Tier 2 models.

The first data set used for Tier 3 training data is a large diverse bot data set that was annotated by detecting 15 digit random alpha-numeric strings as indicated in [12] (a data annotation method using a Tier 0 model). This method provided 1.7 million annotated bot accounts. From this data we built network metrics on 6874 of these accounts. The second data set is from the Debot bot detection system [21] which includes bots that were found due to correlated activity. Using the Debot API, our team extracted 6949 of these accounts, from which we built network metrics on 5939 accounts. Additionally, we used the bot data manually annotated by Cresci et al. in 2015 [23] and again in 2017 [24].

In the results section we will discuss several other data sets that were used to train our Tier 1 and Tier 2 models. These include the annotated data our team captured in a bot attack on the NATO and the Digital Forensic Labs [11]. This data will be referred to as NATO in the results. We also used the suspended Russian bot data set that Twitter released in October 2018 [62]. This data set primarily contains bot/cyborg/troll activity generated by the Russian Internet Research Agency (IRA) during the 2016 US National Elections. In our results sections, this data set is referred to as the IRA data. Finally, we used a large data set of suspended accounts. To acquire this data, our team streamed the 1% Twitter Sample for 7 months, and

Table 3 Data description

Training data	Description	Tier1	Tier2	Tier3
Cresci 2017	Manually annotated by Cresci et al. in 2017 [24]	X	X	X
Cresci 2015	Manually annotated by Cresci et al. in 2015 [23]	X	X	X
Debot data	Accounts labeled as bots by the Debot bot detection system [21]	X	X	X
NATO	Data our team captured in a bot attack on the Digital Forensic Labs and NATO [11]	X	X	
Suspended accounts	These are accounts that were suspended by Twitter	X		
Random string accounts	Accounts with 15 digit random alpha-numeric strings as screen names [12]	X	X	X
IRA data	Suspended Russian bot dataset that Twitter released in October 2018	X		
Combined data	Combination of data listed above	X		

then went back to discover which of the accounts had been suspended. A similar data collection technique was used by Thomas et al. in [61].

The IRA and *suspended* data sets were only used for Tier 1, since timeline and followers were not available for Tier 2 and Tier 3. For the NATO accounts, 96% of the accounts in this dataset have been suspended. We were able to collect sufficient data for Tier 2, but not Tier 3. A summary of each data set is provided in Table 3 and cross walked with the models that it was used with. Note that the Varol data set is not provided here and was not used in our latest bot-hunter models since it is dated and did not perform well.

These data sets contain a wide variety of bots. The Varol data set was founded on the original 2011 Caverlee [50] Honey Pot data, but was supplemented with manual annotations (we leveraged only the manually annotated data). The Cresci data contains both traditional spambots (largely commercial spambots) as well as social spambots (both commercial and political). The random string data contains a large variety of bots ranging from political bots focused on the Middle East to hobby bots focused on Japanese Anime. The Debot data is also fairly diverse, with the one unifying feature that they are all have content and timing correlated with other accounts. The differences in these bots are demonstrated in the t-Distributed Stochastic Neighbor Embedding (t-SNE) dimensionality reduction that we conducted on 2000 randomly sampled accounts from the combined data set (see Fig. 4). Here we see that the Debot Data appears to be separate and different from the Varol, Cresci, and Random String data, which appear to be more uniformly distributed in this 2 dimensional representation of the data.

Fig. 4 t-SNE dimentsionality reduction of Tier 3 feature space (by bot dataset)

In order to train a model, we also needed accounts annotated as *human*. We used the Twitter Streaming API to collect a sample of *normal* Twitter data, intentionally collecting both weekend and weekday data. This provided 149,372 accounts to tag as *human* Twitter accounts. Of these accounts, we were able to collect/measure network metrics on 7614 accounts.

Past research has estimated that 5–8% of twitter accounts are automated [63]. If this is true, then we mis-labeled a small amount of our accounts as *human*. We believe this is acceptable noise in the data, but will limit the performance of supervised machine learning models.

Many other research efforts attempt to annotate human accounts. We chose not to do this because, in the process, these efforts create a biased sample of Twitter, heavily skewed toward average users and under sampling Celebrity, Organizational, Political, Commercial, and other accounts that make up a sizable portion of Twitter discourse. We want our 'human' annotated data to match all non-bot accounts, without biasing it towards any part of this space. Our classification is binary, and the model will be forced to classify all accounts, even those that are under-represented in training data. Our approach therefore attempts to create a truly random sample of Twitter, at the cost of having some bot accounts labeled as 'human'.

4 Feature Engineering

In this section we will introduce our feature engineering for user, content, temporal, and network features. We extracted features from Tier 0 through Tier 3, with a focus on measuring the importance of features extracted from Tier 3. The table of proposed features is provided in Table 4. All new features (beyond the features we presented in [10]) are in bold, and from our research most of these have not been used with an ego-network collected with snowball sampling.

Note that our Tiered approach is cumulative, meaning Tier 3 feature space includes features from Tier 0, Tier 1, and Tier 2. The Tier 3 model therefore includes the Tier 2 network features created by building an entity (mention, hashtag, and URL) co-mention network based only on the user's time-line (last 200 tweets). These Tier 2 network features are distinguished in our results section by the *entity* prefix.

Table 4 Features by data collection tier (new features not presented in [10] highlighted in bold)

Source	User attributes	Network attributes	Content	Timing
User object (Tier 1)	Screen name length	Number of friends	Is last status retweet?	Account age
	Default profile image?	Number of followers	Same language?	Avg tweets per day
	Default profile image?	Number of followers	Same language?	Avg tweets per day
	Entropy screen name	Number of favorites	Hashtags last status	
	Has location?		Mentions last status	
	Total tweets		Last status sensitive?	
	Source (binned)		'bot' reference?	
Timeline (Tier 2)		Number nodes of E	Mean/max mentions	Entropy of inter-arrival
		Number edges	Mean/max hash	Max tweets hour
		Density	Number of languages	Max tweets per day
		Components	Fraction retweets	Max tweets per month
		Largest compo		
		Degree/between centrality		

(continued)

Table 4 (continued)

Source	User attributes	Network attributes	Content	Timing
Snowball sample (Tier 3)	**% w/ default image**	# of bot friends	**# of languages**	**Mean tweets/min**
	Median # tweets	Number of nodes	**Mean emoji per tweet**	**Mean tweets/hour**
	Mean age	Number of links	**Mean mention per tweet**	**Mean tweets/day**
	% w/ description	Density	**Mean hash per tweet**	**% don't sleep**
	% many likes and Few followers	Number of isolates	**% retweets**	
		Number of dyad isolates	**Mean jaccard similarity**	
		Number of triad isolates	**Mean cosine similarity**	
		Number of components >4		
		Clustering coefficient		
		Transitivity		
		Reciprocity		
		Degree centrality		
		K-betweenness centrality		
		Mean eigen centrality		
		Number of simmelian ties		
		Number of Louvain groups		
		Size of largest Louvain group		
		Ego effective size		
		Full triadic census		
		Median followers		
		Median friends		

We hypothesize that the network metrics for human conversations will have different distributions than those made by bot accounts. We also believe that these differences would provide increased performance in traditional machine detection algorithms.

We have not found research that has built a snowball sampling network for bot detection, and believe that all of the Snowball Sampling ego network features in our model are novel. To collect these at scale, our team built a Python package that wrapped around the *networkx* package [39]. We leveraged known network metrics, which are provided in Table 4 with references.

4.1   Network Features

We constructed an ego network from the data collected from snowball sampling, extracting metrics from this network in an effort to develop robust features for bot detection. As discussed earlier, this network consisted of the conversation of the account in question and up to 250 of their followers. All nodes were Twitter accounts, and links were means of directed communication in the Twitter ecosystem (retweet, mention, reply). From this network we developed basic network metrics, component level statistics, centrality metrics, triadic relationship metrics, and clustering related metrics. The basic network metrics are widely used and listed in Table 4. The other categories of metrics are described below.

Given that we did not include the *following* link in our network construction, these networks were not fully connected. As seen in Fig. 5, information from these disconnected components could be valuable in distinguishing real human networks from networks dominated by bots. Our features therefore contain multiple metrics measuring number and size of network components.

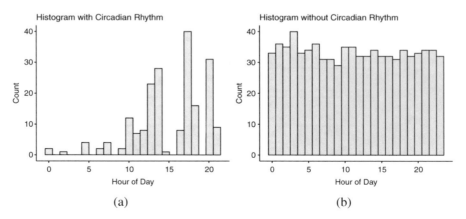

Fig. 5 Differences between a human and bot 24 h circadian rhythms. (**a**) Human *circadian rhythms*. (**b**) Bot *without* circadian rhythms

We included several network centrality metrics in our feature space, and found that they were routinely strong bot predictors. These metrics included mean degree centrality, mean eigenvector centrality, and mean K-betweenness centrality where $K = \min(500, N_{nodes})$.

In addition to analyzing the components, we also computed Louvain grouping and developed metrics based on these groups. We chose the Louvain grouping algorithm given its proven performance on larger data sets. Having computed the Louvain groups, we included metrics such as number of groups and size of largest Louvain group.

Given the importance of triadic relationships in social networks discussed above, we have included several features based on these relationships. These include a full triadic census, number of Simmelian ties, and the clustering coefficient. Calculation of Simmelian ties [46] was not available in the *networkX* package. Our team therefore created a Python implementation of Dekker's version [26] of the original algorithm [46].

4.2   Content Features

We felt we could leverage the large amount of content available from the snowball sample to develop predictive features. This was not done in [10], and was added in recent version of the *bot-hunter* framework.

These features include the number of languages used in the network, as well as some key summary statistics on entities, including mean emojiis, mentions, and hashtags per tweet, as well as the percentage of retweets.

We also wanted to have several measures of similarity of text between the various communicators in the network. This search for similarity measures was motivated by the fact that many bot networks post very similar or conversely very diverse content, and we felt that these measures of similarity may be distinguishing.

To compute similarity, tweet content in the network was aggregated by user. Once aggregated, the content was cleaned and parsed (cleaning included conversion to lower case and removal of punctuation). We did not remove stopwords. The parsed data was then converted to a document term matrix with raw counts (we chose not to normalize the data since the variance on tweet length is artificially constrained to 280 characters). The document term matrix was then used to compute both the Jaccard and Cosine Similarity, which were used as features.

4.3   User Features

The newest version of the Tier 3 classifier also includes several aggregate user attributes that were not leveraged in earlier versions. While many of these are self explanatory, we did want to describe two novel metrics that have not been used before.

Recently, several experts in online disinformation have highlighted how recent online bots seem to produce tweets that are far more popular than the account itself [56]. This phenomena is the result of accounts in large bot-nets that create messages that are then *pushed* by the entire network, resulting in reach that far exceeds expectations given its modest beginning.

To find this phenomena, we devised a simple heuristic that determines if any original (non-retweet) tweet is more popular than its account. This heuristic is defined as:

$$P_{user} = retweets > 2 \times max(followers, friends)$$

where the Boolean measure for a user is defined as $True$ if any tweet receives two time more retweets than the highest value of its in-degree or out-degree. This metric is leveraged in two new features, one at the user level (Tier 2) and one at the Network Level (Tier 3). The user level flags the user if any tweet is flagged as True, and the network metric measures the fraction of tweets produced by the network that are flagged by this heuristic.

4.4   Timing Features

Like user features, most of the temporal features listed in Table 4 are self explanatory. We did develop a heuristic method that measures whether or not an account has daily rhythms. Most human users will have surges in activity based on their daily routines, and will have a measurable drop in activity that aligns with their sleep activity. Bots, on the other hand, do not require these circadian rhythms, and some bots are programmed to produce content spread uniformly across the hours of the day. We developed the heuristic described below to flag these accounts.

To measure whether an account has human circadian rhythms, we first aggregate their tweets by hour of day after ensuring that the account has produced enough data (at least 50 tweets). Given there is sufficient data, we next determine whether this hourly distribution is uniformly distributed by normalizing it and conducting the Kolmogorov-Smirnov non-parametric test for uniformity. A p-value greater than 0.5 provides strong evidence of non-human circadian rhythms.

It is important to note that, while some bots exhibit this lack of circadian rhythm, it only takes a few lines of code for a bot manipulator to give a more realistic temporal pattern. Nonetheless, this remains a strong indicator of bot activity.

5   Modeling

As indicated above, all feature engineering was conducted in Python using several custom Python packages that were developed for the *bot-hunter* framework. These packages build the feature space for Tier 1, Tier2, and Tier 3 models, which is then trained using the steps outlined below.

Table 5 Comparing algorithms for Tier 3 Bot detection

Model	Accuracy	Precision	Recall	AUC	F1
Naïve Bayes	0.562	0.541	0.864	0.563	0.665
Decision tree	0.950	0.949	0.952	0.950	0.951
SVM	0.952	0.969	0.933	0.952	0.952
Logistic regression	0.951	0.940	0.965	0.983	0.952
Random forrest	0.955	0.955	0.956	0.986	0.956

Table 6 Table of results for combined data (Tier 3)

Tier	Accuracy	F1	Precision	Recall	ROC AUC
Tier 1	0.7964	0.7729	0.8677	0.6969	0.8680
Tier 2	0.8335	0.8181	0.8970	0.7522	0.9179
Tier 3	0.8577	0.8478	0.9042	0.7983	0.9410

For training all data sets, human data was sampled so that the classes were balanced. The random forest algorithm was used because of its superior performance on Tier 1 data [11] and its use in other bot detection algorithms [63]. In Table 5 we revisit model comparison in order to verify that the random forest model is still appropriate for Tier 3 feature space. We see that random forest still provides superior performance, and in general is not as computationally expensive as some of the other models. Training, evaluation, and testing were conducted in the *scikit-learn* Python package [57]. Tuning of the Random Forest algorithm was conducted through random search of parameter options while using three fold cross-validation.

The *bot-hunter* behavior returns both a binary classification and an estimate of probability. The estimate of probability is provided by the Random Forest classifier by measuring the proportion of votes by trees in the ensemble. The binary classification result is evaluated by classifying accounts based on a probability threshold of 0.5. The binary classification feature of the results allows researchers to have a consistent threshold to compare results, while the probability allows users to tune a threshold for a given use case.

6   Results

After building the network metrics for all bot data sets as well as the annotated *human* data, we built and evaluated Random Forest models for each of the data sets. Training, evaluating, and testing were conducted at Tier 1, Tier 2, and Tier 3 where possible. We evaluated in-sample performance with 10 fold cross-validation measuring multiple evaluation metrics, which are provided in Table 6 and Fig. 6.

From the results presented in Table 6 and Fig. 6, we see that Tier 1 models continue to provide solid performance, even with basic features extracted from the user profile and last status. We also observe improvement between Tier 1

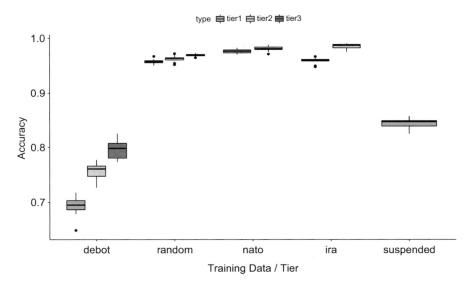

Fig. 6 Results by training data and by *Tier*

and Tier 2 and between Tier 2 and Tier 3 for all models. Using a combined data model we found that the Tier 2 improvement over Tier 1 is statistically significant (p−value $= 1.303e − 10$), as is the Tier 3 improvement over Tier 2 (p−value $= 1.101e − 06$). In Fig. 6 we also see that the Random, NATO, and IRA data provide the highest in sample cross validation performance, while models trained on Debot Data and Suspended data offer lower in data cross validation performance. This likely indicates a wider variety of bot types in the Debot and Suspended data.

Further, in Fig. 7 we see the top features for all Tier 1, Tier 2, and Tier 3 models in the *bot-hunter* suite of tools. These figures represent the percentage that each feature contributed to the model predictions. We see that network features provide strong features in the model. This demonstrates that these values, while tedious to collect, transform, and model, provide strong predictive features that are difficult for bot *puppet master* to manipulate. In these data sets network centrality, network connection, network timing, and network content all provide predictive value.

7   Evaluating Against State of the Art

Given that this is the last Tier of the *bot-hunter* suite of tools, we wanted to evaluate the models as well as various training data that is available. We also wanted to compare the models in the *bot-hunter* suite of tools to existing models, namely the Botometer and Debot models. To do this, we set out to find a test that wasn't biased

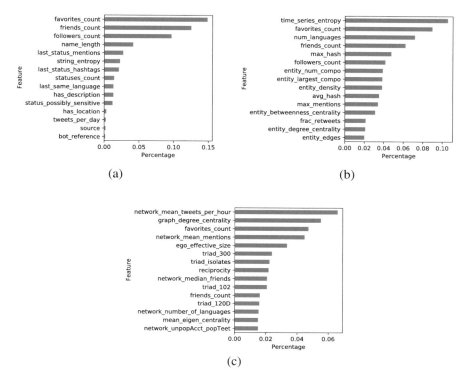

Fig. 7 Comparison of top features for all three Tiers of *bot-hunter*. (**a**) Tier 1 Top Predictive Features. (**b**) Tier 2 Top Predictive Features. (**c**) Tier 3 Top Predictive Features

toward any given model, meaning the test data could not be derived from the training data of any of the models being compared.

To find an unbiased data set, we manually annotated 337 bot accounts. To do this, we started by manually finding several seed bots related to the Swedish elections, separate Russian propaganda bots, and bots found in Middle East conversations. We then manually snowballed out on the followers and followers of followers, manually identifying additional bots. In this evaluation we leveraged the visualizations and metrics provided in the TruthNest tool to aid in making our determination. The TruthNest Tool originally was an EU-funded Reveal project developed to evaluate Twitter accounts for automated activity. While this tool was not evaluated in our test, it was used to assist in labeling bot accounts. TruthNest has instituted a paywall since our use of it. Human users were sampled from the Twitter stream and manually verified. The test data was balanced (337 bots, 337 users).

In evaluating our Tier1, Tier2, and Tier3 models, we also wanted to evaluate which training data and model combination generalizes to new data. Our models were trained on the data and at the tiers described in Table 1. All *bot-hunter* and Botometer thresholds were set at 0.5. F1 performance for all models is provided in Fig. 8 and detailed results are provided in Table 7.

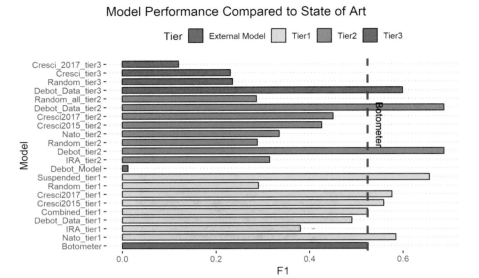

Fig. 8 Results by training data and by *Tier*

Table 7 Detailed results by Tier and training data

Tier	Training data	F1	Accuracy	Precision	Recall	ROC-AUC	TN	FP	FN	TP
Botometer model		0.524	0.657	0.858	0.377	0.587	256	55	200	108
Debot model		0.012	0.502	1.000	0.006	0.503	336	0	335	2
Tier1	NATO	0.584	0.634	0.678	0.513	0.635	254	82	164	173
Tier1	IRA	0.380	0.597	0.830	0.246	0.598	319	17	254	83
Tier1	Combined	0.524	0.657	0.858	0.377	0.657	315	21	210	127
Tier1	Cresci2015	0.559	0.404	0.444	0.754	0.404	18	318	83	254
Tier1	Cresci2017	0.576	0.419	0.454	0.789	0.418	16	320	71	266
Tier1	Debot	0.490	0.527	0.533	0.454	0.528	202	134	184	153
Tier1	Random	0.291	0.572	0.855	0.175	0.573	326	10	278	59
Tier1	Suspended	0.656	0.713	0.821	0.546	0.713	296	40	153	184
Tier2	IRA	0.315	0.567	0.903	0.191	0.584	305	7	276	65
Tier2	Random	0.288	0.547	0.800	0.176	0.564	297	15	281	60
Tier2	NATO	0.335	0.574	0.909	0.205	0.591	305	7	271	70
Tier2	Cresci2015	0.426	0.596	0.824	0.287	0.610	291	21	243	98
Tier2	Cresci2017	0.451	0.600	0.799	0.314	0.614	285	27	234	107
Tier2	Debot	0.687	0.675	0.691	0.683	0.675	208	104	108	233
Tier2	Random	0.286	0.550	0.831	0.173	0.567	300	12	282	59
Tier3	Debot	0.599	0.674	0.837	0.466	0.683	281	31	182	159
Tier3	Random	0.236	0.533	0.810	0.138	0.551	301	11	294	47
Tier3	Cresci2015	0.231	0.541	0.918	0.132	0.560	308	4	296	45
Tier3	Cresci2017	0.120	0.507	0.880	0.065	0.527	309	3	319	22

In these results we first see Botometer demonstrates consistent solid performance in predicting new bots across all metrics. The Debot algorithm provides high precision but extremely low recall, resulting in a low F1 score overall. The value of the Debot algorithm may indirectly lie in the data that it produces. Note that *bot-hunter* algorithms trained on Debot data performed well at all three Tiers, meaning that the Debot algorithm for finding correlated accounts produces great labeled data for other supervised bot detection endeavors.

For the bot-hunter family of models, we see that Tier 1 consistently performs well and seems to generalize to new data better than Tier 2 and Tier 3. Tier 2 still has high performance, given its ability to identify anomalies in content and in temporal statistics. Across the data sets, Tier 1 has a higher mean Accuracy and ROC AUC than Tier 1. Tier 3 has very high precision but low recall. It therefore produces predictions that are more reliable, but fails to find a large portion of the bots in the data. Additionally, this model may become increasingly important in identifying sophisticated emerging bots.

As we look at the various training data used for training these models, we see that the models trained on suspended accounts or on data produced by the Debot model had the highest performance. As indicated earlier, this is likely due to these data sets containing a wide variety of bot "genres." We also see that the NATO data captured in the deliberate attack against NATO and the DFR labs continues to provide strong performance across all metrics. We found that few of the annotated data sets released by other researchers provided strong performance, especially when considering accuracy and ROC-AUC metrics. The Cresci data (both 2015 and 2017) appears to have high recall but low precision, with many false negatives. The models trained on the random string data also have low accuracy and ROC-AUC metrics, in this case caused by high precision but low recall. These random string accounts probably represent a limited band in the spectrum of bot types, and therefore do not generalize well to new data and different bot types.

The Venn Diagram of predicted bots is provided in Fig. 9a. This diagram shows the overlap of the predicted bots, but does not provide any information on predicted humans. We see significant overlap for all three models. We also notice that the Tier 2 model predicted the most accounts (330 accounts), while Tier 1 predicted 260 accounts and Tier 3 predicted 183 bot accounts. The 95 accounts in the intersection contain 20 false positives (78.9% precision).

The Venn Diagram of predicted bots for Tier 1 and 2 compared to the *real* labeled bots is provided in Fig. 9b. This shows that Tier 2 is adding something to Tier 1, finding 94 additional accounts while only missing 21 of the accounts that Tier 1 found.

Figure 9c provides an *upset* visualization to fully explore the intersection of sets. This visualization demonstrates that our largest intersection is the intersection of all four sets. We also see in the upset graph the Tier 1 and in Particular Tier 2 is important to the prediction success, thought Tier 3 is also able to find 32 accounts that neither Tier 1 or 2 could find. These visualizations illustrate the importance of having a tool-box of models that can be used for predicting bots in any given scenario.

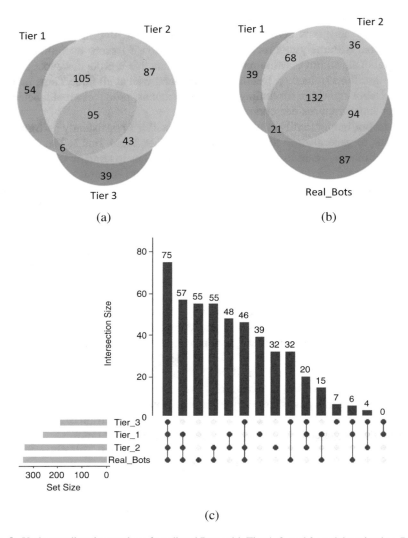

Fig. 9 Understanding the overlap of predicted Bots with Tier 1, 2, and 3 models trained on Debot data. (**a**) Predicted Bots (Tier 1, 2, and 3) . (**b**) Predicted Bots (Tier 1 and 2) with Real Labeled Bots. (**c**) Upset Plot with Predicted Bots (Tier 1, 2, and 3) and Real Labeled Bots

While we believe this evaluation is informative, there are several limitations in our evaluation method. We aknowledge that we were not able to completely remove bias, given that the mental heuristics we used to manually annotate accounts may have unintentionally mirrored the bot-hunter algorithms. Additionally, we acknowledge that the test set is still modest in size and, while somewhat diverse, does not represent the full spectrum of bot types. Finally, we acknowledge that any given model may perform better if the threshold is tuned for a given data set. Even with these limitations, we believe this test and evaluation is informative for our team and for the greater community.

7.1 Evaluating Bot Classification Thresholds

The random forest model used in the bot-hunter suite of tools (and Botometer) provides a probability estimate rather than just a label. This allows researchers to estimate how strong a given prediction is. Every use case will require the analyst to determine the best threshold for establishing whether or not an account is likely a bot. To evaluate the best threshold for a given data set, a research team should explore several thresholds, each time sampling 50–100 accounts and manually labeling them to estimate a rate of true/false positives, true/false negatives. If possible they should attempt to construct a precision recall curve and/or ROC Curve, as demonstrated in Fig. 10 using the Suspended, NATO, and Botometer models. Note that recall is always monotonically decreasing, but precision is not required to monotonically increase.

As seen in Fig. 10, we generally recommend bot-hunter thresholds between 0.6 and 0.8. The exact choice in this range will need to be made by the research team, and is dependent on the data as well as the team's prioritization of precision vs. recall.

8   Applying Bot Detection to Swedish Election

Having completed the bot-hunter suite of tools, we wanted to leverage this toolbox in analyzing a stream of data from the 2018 Swedish elections. This is done as a case study to illustrate that bot-detection is not a "turn-key" solution, and also to provide practitioners with an example of an open source intelligence workflow.

Sweden held national elections on 9 September 2018 for its equivalent of a Parliament, known as the Riksdag. Swedish elections have historically lacked much drama or suspense, with the center-left Social Democrat Party dominating politics since 1914. In the 2018 election, however, their dominance was challenged by various nationalistic factions that capitalized on anti-immigrant sentiment.

Some of the political discourse surrounding the election transpired on Twitter, as seen in many recent national elections across the world. As this discourse grew, multiple researchers and news agencies saw rising disinformation and associated bot activity [58]. Simultaneously, the Swedish Defence Research Agency reported increased bot activity, primarily supporting right leaning, nationalistic, and anti-immigrant views [60].

As these bots grew in activity in this marketplace of beliefs and ideas, our team began collecting and analyzing streams from this discourse. To collect Twitter

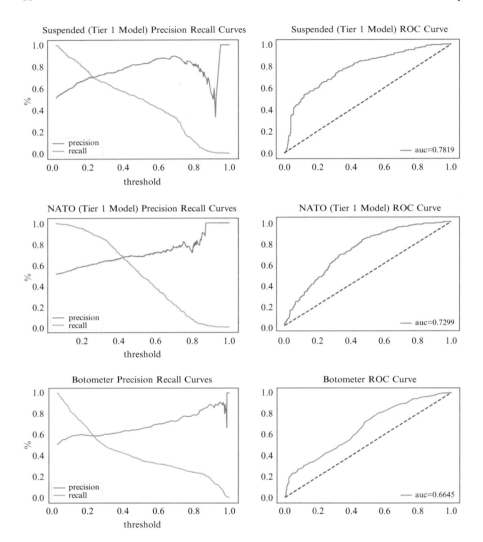

Fig. 10 Using precision-recall curves and ROC curves to determine threshold

data around the Swedish National elections discourse, our team leveraged a spiral collection methodology, starting with content and geographic streaming, and then 'spiraling' into more thorough data collection around the important parts of the discussion. All collection was done through the Twitter Streaming and REST API's using the Tweepy Python Package.

We started by identifying Swedish political hashtags through open source research, eventually identifying #svpol, #Val2018, #feministisktInitiativ, #migpol, #valet2018, #SD2018, #AfS2018, and #MEDval18. These hashtags were not selected because they cover the full spectrum of Swedish politics, but rather because there was open source reporting of some bot campaigns using these hashtags. We started collecting on these hashtags using both the Streaming and REST API's (the streaming API allows us to easily collect going forward while the REST API allows us to retroactively collect past data). Simultaneously we collected data that was 'geo-associated' with the Scandinavian peninsula, using a bounding box search method.

As we began to collect content and geo-referenced data, we monitored other trending hashtags and added them to the collection query. After launching the exploratory data analysis discussed below, we would also collect users friend and follower relationships as well as user historical timelines for accounts of interest. This continual return to the Twitter API creates the spiral nature of our collection process.

For the Swedish Election Event we collected 661,317 tweets produced by 88,807 unique users. This creates a political *conversation* that contains 104,216 nodes, 404,244 links with a density of 0.000037.

For bot detection in the Swedish Election stream our team found that a 65% probability was appropriate. Given that we were performing this evaluation on 104,216 nodes, we used the Tier 1 model. This model is our best model for getting an accurate prediction on high volume of accounts.

Note that we usually conduct other data enhancement as well, including sentiment analysis with NetMapper as well as geo-inference based on [44]. All enrichments are made available in easy formats that allow tools to merge them with existing event data.

8.1   Exploratory Data Analysis

Our exploratory data analysis focuses on narratives, time, place, groups, and individuals. Our analysis typically starts with some type of temporal analysis. This allows us to see distributions over time. We try to look at overall temporal distribution, bot activity over time, as well as changing narratives over time (Fig. 11).

Our exploration of content and narratives starts with analysis of words and hashtags across the entire corpus, and then we explore narratives associated with topic groups (these are groups that talk about the same thing but may not be connected in the social network or conversational network) and social network group (these are groups that are connected, but may not talk about the same thing). We leverage latent dirichlet allocation [15] for topic group analysis, and content analysis by Louvain group [16] as a way to "triage" network groups. Table 8 provides the top 8 words by Louvain Group for the Swedish elections. In this we already start to see groups that are focused on immigration, particularly immigration from Muslim countries. We also see at least one group that is mixing conversation about religious beliefs with political discourse. Finally and just as

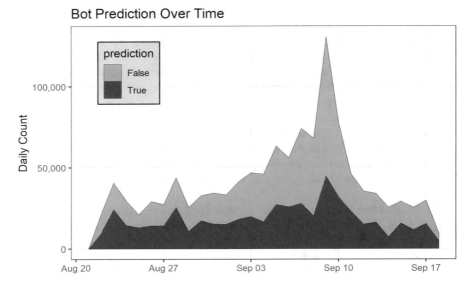

Fig. 11 Bots as a proportion of total volume over time

important, "triaging" the data like this allows us to identify groups like Group 0 that don't appear to have any topics of interest.

Network analysis of groups and individuals is done almost exclusively in the ORA Network Analysis Tool. We typically start by visualizing a reduced conversational network. Nodes in this network represent Twitter accounts, and links represent a conversational action in the Twitter ecosystem (reply, retweet, mention). These network are typically too large to visualize, so we reduce the network by taking the K-core so that we have the core 15,000–20,000 nodes. Once this is done, we color the network by *bot* or *human*, by language, and by Louvain grouping (see Fig. 12). This coloration helps us better understand the groups and their relationship to each other. Finally, we reduce the network to only include reciprocal links. This usually reduces the network significantly, and in Twitter provides the best proxy for a true social network.

We then explore the influential accounts and influential bots in the network. The ORA Network analysis tool provides several reports that analyze nodes by a variety of centrality measures, and assists translating their role in the network. For the Swedish network, we found several bots with high *betweenness*, indicating that these bots were influential in that they connect individuals and groups. With further exploration, it appeared that these bots, in connection with other accounts, were trying to bridge several communities with nationalistic and anti-immigrant groups and narratives.

We leverage the *bot-hunter* Tier 2 and Tier 3 models during this phase of analysis. As we identify influential accounts, we check them in a Tier 2/3 *bot-hunter* web application that allows us to thoroughly explore the account and conduct a more

Table 8 Content analysis by Louvain group

Group	# Tweets	# Nodes	Top 8 words by Louvain group			
0	15,708	4675	Video 2018	Gillade fortnite	Lade world	Spellista part
1	31,059	5688	Country n	Voters number	Refugees capita	82 reported
2	102,146	14,538	Sweden Swedish	Election results	epp left	sd poll
3	306,352	17,600	m6aubkudbg sverige	Jesus gud	Kristus namn	Varnar fader
4	8353	3137	Sweden amp	Swedish vote	Muslim democrats	Election gang
5	40,585	9110	Sverige valet	sd jimmie	Svenska år	åkesson svt
7	82,708	12,300	sd valet	Sverige jimmie	Rösta parti	åkesson val
8	17,675	4000	sd politik	Friend claeson	American tånkt	Rösta frågar
9	7144	5217	Sverige stefan	Löfven kristersson	sd amp	Moderaterna rösta
10	7569	5214	Sverige afs	Riks Sweden	sd svenska	Alternativ hahne

accurate Tier 2 or Tier 3 bot prediction. These applications also allow us to explore in depth visualizations of the activity of the account.

Bot-detection is therefore a part of the overall open source intelligence workflow, trying to identify relevant information about how the world works to inform decision maker situational understanding and decisive action. In this case, our research validated research of large bot activity within the Swedish political discourse on Twitter and provided identification of narratives (primarily nationalistic and anti-immigrant, anti-Muslim, and some anti-Semitism). We were also able to identify influential accounts that were attempting to connect individuals and online communities with extremist content. This type of information informs leaders of current dis-information strategies allowing them to better prepare their government and their populace for similar disinformation campaigns in their country.

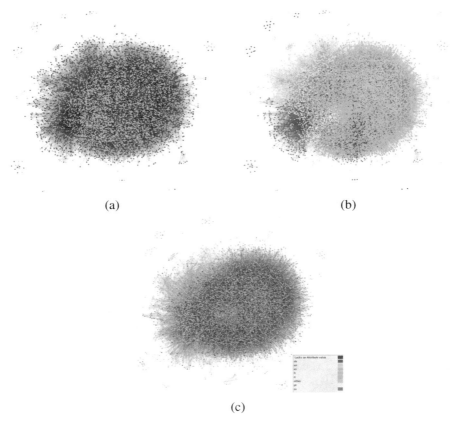

(a)   (b)

(c)

Fig. 12 Exploring the Twitter conversational network surrounding online discourse on Swedish politics. (**a**) Bots (red) in conversation. (**b**) Louvain Groups. (**c**) Language Distribution in Network

9   Conclusion and Future Work

In our pursuit of a multi-model bot detection toolbox, this paper builds on past research by adding a model that leverages a feature space extracted from 50,000+ entities collected with single seed snowball sampling. This model is developed for high accuracy but low volume applications. Our research shows that supervised machine learning models are able to leverage these rich structural, content, and temporal features associated with the target ego-network to increase model precision. Additionally, these network features offer an approach for modeling and detecting bot behavior that is difficult for bot *puppet-masters* to manipulate and evade.

Our evaluation of the bot-hunter suite of tools demonstrates that these models provide performance equivalent to or better than the state of the art. The Tier 1 model in particular is valuable to the community because it is accurate and can scale to large data (meaning researchers aren't required to sample their data). Additionally, because the Tier 1 model was designed to predict existing data, there isn't a requirement to return to the Twitter API to re-collect account data. This also means that it can be used to predict existing data sets that contain suspended or otherwise missing accounts.

Our analysis of Swedish political discourse on Twitter illustrates how bot-detection tools can support a typical open source intelligence workflow. The bot-hunter suite provides a way to enrich the data which can then be imported into other analysis tools for visualization and further analysis. Bot detection is not a "turn-key" solution, and does require some work to set the right parameters, particularly the appropriate threshold level.

Future work will focus on creating a labeling methodology that will allow us to better characterize bot accounts and the various methods they employ. Binary prediction assists in understanding fake versus real, but does not help us in triaging the hundreds of thousands of bot accounts that exist. Some spam content, others intimidate users. Developing heuristics to label these methods and attributes is essential for characterizing these accounts and the disinformation campaigns they propagate.

Acknowledgements This work was supported in part by the Office of Naval Research (ONR) Multidisciplinary University Research Initiative Award N000140811186 and Award N000141812108, the Army Research Laboratory Award W911NF1610049, Defense Threat Reductions Agency Award HDTRA11010102, and the Center for Computational Analysis of Social and Organization Systems (CASOS). The views and conclusions contained in this document are those of the authors and should not be interpreted as representing the official policies, either expressed or implied, of the ONR, ARL, DTRA, or the U.S. government.

References

1. K.S. Adewole, N.B. Anuar, A. Kamsin, K.D. Varathan, S.A. Razak, Malicious accounts: dark of the social networks. J. Netw. Comput. Appl. **79**, 41–67 (2017)
2. L.M. Aiello, M. Deplano, R. Schifanella, G. Ruffo, People are strange when you're a stranger: Impact and influence of bots on social networks. Links **697**(483,151), 1–566 (2012)
3. A. Almaatouq, E. Shmueli, M. Nouh, A. Alabdulkareem, V.K. Singh, M. Alsaleh, A. Alarifi, A. Alfaris, et al., If it looks like a spammer and behaves like a spammer, it must be a spammer: analysis and detection of microblogging spam accounts. Int. J. Inf. Secur. **15**(5), 475–491 (2016)
4. A. Ananthalakshmi, *Ahead of Malaysian Polls, Bots Flood Twitter with Pro-government...* (2018)
5. D.A Bader, S. Kintali, K. Madduri, M. Mihail, Approximating betweenness centrality, in *International Workshop on Algorithms and Models for the Web-Graph* (Springer, Berlin, 2007), pp. 124–137
6. A. Bavelas, A mathematical model for group structures. Hum. Organ. **7**(3), 16–30 (1948)
7. A.I. Bawaba, The Loop, in *Thousands of Twitter Bots are Attempting to Silence Reporting on Yemen* (2017)

8. M. Benigni, K.M. Carley, From tweets to intelligence: understanding the islamic jihad supporting community on twitter, in *Proceedings of Social, Cultural, and Behavioral Modeling: 9th International Conference, SBP-BRiMS 2016, Washington, DC, USA, June 28-July 1, 2016* (Springer, Berlin, 2016), pp. 346–355

9. M.C. Benigni, K. Joseph, K.M. Carley, Online extremism and the communities that sustain it: detecting the ISIS supporting community on twitter. PLoS One **12**(12), e0181405 (2017)

10. D. Beskow, K.M. Carley, Bot conversations are different: Leveraging network metrics for bot detection in twitter, in *2018 International Conference on Advances in Social Networks Analysis and Mining (ASONAM)* (IEEE, Piscataway, 2018), pp. 176–183

11. D. Beskow, K.M. Carley, Introducing bothunter: A tiered approach to detection and characterizing automated activity on twitter, in *International Conference on Social Computing, Behavioral-Cultural Modeling and Prediction and Behavior Representation in Modeling and Simulation*, ed. by H. Bisgin, A. Hyder, C. Dancy, R. Thomson (Springer, Berlin, 2018)

12. D. Beskow, K.M. Carley, Using random string classification to filter and annotate automated accounts, in *International Conference on Social Computing, Behavioral-Cultural Modeling and Prediction and Behavior Representation in Modeling and Simulation*, ed. by H. Bisgin, A. Hyder, C. Dancy, R. Thomson (Springer, Berlin, 2018)

13. A. Bessi, E. Ferrara, *Social Bots Distort the 2016 US Presidential Election Online Discussion* (2016)

14. S.Y. Bhat, M. Abulaish, Community-based features for identifying spammers in online social networks, in *2013 IEEE/ACM International Conference on Advances in Social Networks Analysis and Mining (ASONAM)* (IEEE, Piscataway, 2013), pp. 100–107

15. D.M. Blei, A.Y. Ng, M.I. Jordan, Latent dirichlet allocation. J. Mach. Learn. Res. **3**(Jan), 993–1022 (2003)

16. V.D. Blondel, J.-L. Guillaume, R. Lambiotte, E. Lefebvre, Fast unfolding of communities in large networks. J. Stat. Mech: Theory Exp. **2008**(10), P10008 (2008)

17. R.S. Burt, *Structural Holes: The Social Structure of Competition* (Harvard University Press, Cambridge, 2009)

18. K.M. Carley, G. Cervone, N. Agarwal, H. Liu, Social cyber-security, in *International Conference on Social Computing, Behavioral-Cultural Modeling and Prediction and Behavior Representation in Modeling and Simulation*, ed. by H. Bisgin, A. Hyder, C. Dancy, R. Thomson (Springer, Berlin, 2018)

19. D. Cartwright, F. Harary, Structural balance: a generalization of Heider's theory. Psychol. Rev. **63**(5), 277 (1956)

20. N. Chavoshi, H. Hamooni, A. Mueen, Debot: Twitter bot detection via warped correlation, in *IEEE International Conference on Data Mining (ICDM)* (2016), pp. 817–822

21. N. Chavoshi, H. Hamooni, A. Mueen, On-demand bot detection and archival system, in *Proceedings of the 26th International Conference on World Wide Web Companion*. International World Wide Web Conferences Steering Committee (2017), pp. 183–187

22. C.-M. Chen, D.J. Guan, Q.-K. Su, Feature set identification for detecting suspicious urls using bayesian classification in social networks. Inform. Sci. **289**, 133–147 (2014)

23. S. Cresci, R. Di Pietro, M. Petrocchi, A. Spognardi, M. Tesconi, Fame for sale: efficient detection of fake twitter followers. Decis. Support. Syst. **80**, 56–71 (2015)

24. S. Cresci, R. Di Pietro, M. Petrocchi, A. Spognardi, M. Tesconi, Social fingerprinting: detection of spambot groups through DNA-inspired behavioral modeling. IEEE Trans. Dependable Secure Comput. **15**(4), 561–576 (2018)

25. C.A. Davis, O. Varol, E. Ferrara, A. Flammini, F. Menczer, Botornot: a system to evaluate social bots, in *Proceedings of the 25th International Conference Companion on World Wide Web*. International World Wide Web Conferences Steering Committee (2016), pp. 273–274

26. D.J. Dekker, Measures of Simmelian Tie Strength, Simmelian Brokerage, and, the Simmelianly Brokered (2006)

27. R.I.M. Dunbar, Coevolution of neocortical size, group size and language in humans. Behav. Brain Sci. **16**(4), 681–694 (1993)

28. E. Ferrara, Measuring social spam and the effect of bots on information diffusion in social media (2017). arXiv preprint:1708.08134
29. S. Fortunato, Community detection in graphs. Phys. Rep. **486**(3–5), 75–174 (2010)
30. O. Frank, Sampling and estimation in large social networks. Soc. Networks **1**(1), 91–101 (1978)
31. L.C. Freeman, Centrality in social networks conceptual clarification. Soc. Networks **1**(3), 215–239 (1978)
32. L.C. Freeman, Centered graphs and the structure of ego networks. Math. Soc. Sci. **3**(3), 291–304 (1982)
33. K. Gani, H. Hacid, R. Skraba, Towards multiple identity detection in social networks, in *Proceedings of the 21st International Conference on World Wide Web* (ACM, New York, 2012), pp. 503–504
34. M. Gjoka, M. Kurant, C.T. Butts, A. Markopoulou, Practical recommendations on crawling online social networks. IEEE J. Sel. Areas Commun. **29**(9), 1872–1892 (2011)
35. A. Glaser, *Russian Bots are Trying to Sow Discord on Twitter After Charlottesville* (2017)
36. L.A. Goodman, Snowball sampling, in *The Annals of Mathematical Statistics* (1961), pp. 148–170
37. M.S. Granovetter, The strength of weak ties, in *Social Networks* (Elsevier, Amsterdam, 1977), pp. 347–367
38. C. Grimme, M. Preuss, L. Adam, H. Trautmann, Social bots: human-like by means of human control? Big Data **5**(4), 279–293 (2017)
39. A. Hagberg, P. Swart, D.S. Chult, *Exploring Network Structure, Dynamics, and Function Using Networkx* (Los Alamos National Lab.(LANL), Los Alamos, 2008). Technical report
40. F. Heider, Attitudes and cognitive organization. J. Psychol. **21**(1), 107–112 (1946)
41. P.W. Holland, S. Leinhardt, Transitivity in structural models of small groups. Compar. Group Stud. **2**(2), 107–124 (1971)
42. P.W. Holland, S. Leinhardt, A method for detecting structure in sociometric data, in *Social Networks* (Elsevier, Amsterdam, 1977), pp. 411–432
43. P.N. Howard, B. Kollanyi, Bots,# strongerin, and# brexit: computational propaganda during the uk-eu referendum, in *Browser Download This Paper* (2016)
44. B. Huang, K.M. Carley, On predicting geolocation of tweets using convolutional neural networks, in *International Conference on Social Computing, Behavioral-Cultural Modeling and Prediction and Behavior Representation in Modeling and Simulation* (Springer, Berlin, 2017), pp. 281–291
45. L. Katz, A new status index derived from sociometric analysis. Psychometrika **18**(1), 39–43 (1953)
46. D. Krackhardt, The ties that torture: Simmelian tie analysis in organizations. Res. Sociol. Organ. **16**(1), 183–210 (1999)
47. S. Kudugunta, E. Ferrara, Deep neural networks for bot detection (2018). arXiv preprint:1802.04289
48. S. Lee, J. Kim, Early filtering of ephemeral malicious accounts on twitter. Comput. Commun. **54**, 48–57 (2014)
49. K. Lee, J. Caverlee, S. Webb, Uncovering social spammers: social honeypots+ machine learning, in *Proceedings of the 33rd International ACM SIGIR Conference on Research and Development in Information Retrieval* (ACM, New York, 2010), pp. 435–442
50. K. Lee, B.D. Eoff, J. Caverlee, Seven months with the devils: a long-term study of content polluters on twitter, in *ICWSM* (2011)
51. D. Liu, B. Mei, J. Chen, Z. Lu, X. Du, Community based spammer detection in social networks, in *International Conference on Web-Age Information Management* (Springer, Berlin, 2015), pp. 554–558
52. C. Lumezanu, N. Feamster, H. Klein, # bias: Measuring the tweeting behavior of propagandists, in *Sixth International AAAI Conference on Weblogs and Social Media* (2012)
53. Z. Miller, B. Dickinson, W. Deitrick, W. Hu, A.H. Wang, Twitter spammer detection using data stream clustering. Inf. Sci. **260**, 64–73 (2014)

54. A. Mislove, M. Marcon, K.P. Gummadi, P. Druschel, B. Bhattacharjee, Measurement and analysis of online social networks, in *Proceedings of the 7th ACM SIGCOMM conference on Internet measurement* (ACM, New York, 2007), pp. 29–42
55. L.M. Neudert, B. Kollanyi, P.N. Howard, *Junk News and Bots During the German Federal Presidency Election: What Were German Voters Sharing Over Twitter?* (2017)
56. B. Nimmo, *#botspot: Twelve Ways to Spot a Bot—dfrlab—medium* (2017). https://medium.com/dfrlab/botspot-twelve-ways-to-spot-a-bot-aedc7d9c110c (Accessed on 11/03/2018).
57. F. Pedregosa, G. Varoquaux, A. Gramfort, V. Michel, B. Thirion, O. Grisel, M. Blondel, P. Prettenhofer, R. Weiss, V. Dubourg, J. Vanderplas, A. Passos, D. Cournapeau, M. Brucher, M. Perrot, and E. Duchesnay, Scikit-learn: machine learning in Python. J. Mach. Learn. Res. **12**, 2825–2830 (2011)
58. J. Stubbs, J. Ahlander, *Exclusive: Right-Wing Sites Swamp Sweden with 'Junk News' in Tight Election Race | Reuters* (2018). https://www.reuters.com/article/us-sweden-election-disinformation-exclus/exclusive-right-wing-sites-swamp-sweden-with-junk-news-in-tight-election-race-idUSKCN1LM0DN (Accessed on 11/20/2018).
59. V.S. Subrahmanian, A. Azaria, S. Durst, V. Kagan, A. Galstyan, K. Lerman, L. Zhu, E. Ferrara, A. Flammini, F. Menczer, The darpa twitter bot challenge. Computer **49**(6), 38–46 (2016)
60. Swedish Defence Reserch Agency, Antalet botar på twitter ökar inför valet—totalförsvarets forskningsinstitut (2018). https://www.foi.se/press--nyheter/nyheter/nyhetsarkiv/2018-08-29-antalet-botar-pa-twitter-okar-infor-valet.html (Accessed on 11/20/2018)
61. K. Thomas, C. Grier, D. Song, V. Paxson, Suspended accounts in retrospect: an analysis of twitter spam, in *Proceedings of the 2011 ACM SIGCOMM conference on Internet measurement conference* (ACM, New York, 2011), pp. 243–258
62. Twitter. *Elections Integrity Data Archive.* https://about.twitter.com/en_us/values/elections-integrity.html#us-elections (Accessed on 03/30/2019)
63. O. Varol, E. Ferrara, C.A. Davis, F. Menczer, A. Flammini, Online human-bot interactions: Detection, estimation, and characterization (2017). arXiv preprint:1703.03107
64. J.-P. Verkamp, M. Gupta, Five incidents, one theme: Twitter spam as a weapon to drown voices of protest, in *Presented as part of the 3rd USENIX Workshop on Free and Open Communications on the Internet* (2013).
65. B. Viswanath, M.A. Bashir, M. Crovella, S. Guha, K.P. Gummadi, B. Krishnamurthy, A. Mislove, Towards detecting anomalous user behavior in online social networks, in *USENIX Security Symposium* (2014), pp. 223–238
66. G. Wang, M. Mohanlal, C. Wilson, X. Wang, M. Metzger, H. Zheng, B.Y. Zhao, Social turing tests: crowdsourcing sybil detection (2012). arXiv preprint:1205.3856
67. H. Yu, M. Kaminsky, P.B. Gibbons, A. Flaxman, Sybilguard: defending against sybil attacks via social networks, in *ACM SIGCOMM Computer Communication Review*, vol. 36 (ACM, New York, 2006), pp. 267–278

Beyond the 'Silk Road': Assessing Illicit Drug Marketplaces on the Public Web

Richard Frank and Alexander Mikhaylov

Abstract Criminals take advantage of internet communications to amplify the impact of their actions and to form international criminal networks. At the same time, vast amounts of information generated by their online activities have become available for analysis. Open source web intelligence is a valuable methodology for understanding and responding to these new global criminal phenomena. Collecting data from websites, social media platforms and online discussion forums enables researchers, investigators and policy-makers to study and to develop appropriate responses to emerging threats. Automated web intelligence tools such as web crawlers can be used to extract relevant information from target websites and to map the threat landscape of criminogenic environments online. For the study presented in this chapter, we used our web-crawling software to download contents of 28 Russian online marketplaces for illicit drugs. Drug names, types, prices, quantities and geographical locations of sales were extracted and mapped to identify drug trafficking hotspots. Findings indicate such marketplaces can operate due to the ability of their clients to pay anonymously with virtual currencies (specifically Bitcoin and Qiwi) and to deliver the drugs through non-contact methods. This type of service is available in all large cities within Russia and provides to the seller with a safer and more anonymous alternative to "street-level" purchases. The method described in this study can be used to investigate and to prioritize online threats according to their location and severity.

Keywords Online drug trafficking · Anonymous payments · Novel psychoactive substances

R. Frank (✉) · A. Mikhaylov
School of Criminology, Simon Fraser University, Burnaby, BC, Canada
e-mail: rfrank@sfu.ca; amikhayl@sfu.ca

© Springer Nature Switzerland AG 2020
M. A. Tayebi et al. (eds.), *Open Source Intelligence and Cyber Crime*, Lecture Notes in Social Networks, https://doi.org/10.1007/978-3-030-41251-7_4

1 Introduction

The proliferation of internet connectivity throughout the world has facilitated the growth of legitimate and illegitimate economies alike. Globalization has made the international geographical borders, and the physical distance between buyers and suppliers, largely irrelevant. The digital economy has allowed consumers to pay bills, make money transfers and purchase goods with unprecedented convenience, whether locally or internationally. Online payments first emerged in the late 1990s in the form e-gold services, owned by Gold & Silver Reserve Inc. (G&SR), which allowed users to open online accounts and purchase gold (and other precious metals) in grams, then transfer them to other accounts as payments [1]. Whether used for legitimate business opportunities, or shady gambling websites and virtual Ponzi schemes, e-gold kickstarted the virtual currency industry [1]. Competitors quickly emerged, and the competition within the e-currency industry forced market actors to innovate and to offer increasingly varied types of exchange services for converting traditional fiat currency into virtual currency for online transfers/payments. The key feature which came to define the success or failure of a virtual currency is the ease of depositing money into an online account [1]. Increased regulatory and law enforcement attention to e-gold and novel digital currencies in the mid-2000s has resulted in traditional banking systems, especially in the United States, distancing themselves from e-currencies [1]. However, this merely meant e-currency companies would operate outside jurisdictions with strict regulations.

E-currencies and mobile payments often avoid anti-money laundering (AML) legislation as virtual currencies are not emitted or controlled by banks and as a result these organizations are not subject to the same regulations [2]. With the advent of e-commerce, the goods and services offered on underground markets have become increasingly diverse. While initially focused on trading in stolen credentials, stolen financial data and hacking, a new sector of the underground economy rose to prominence in 2011 with the creation of Silk Road—the first crypto-market for illicit drugs, with many successors to follow [3]. While a European survey showed that most illicit drug purchases among young people still relied on traditional retail distribution models, the United Nations Office on Drugs and Crime (UNODC) reported 90% of UNODC member states identified the internet as a source of drug trafficking [4]. It is especially true in relation to "novel psychoactive substances" (NPS), also known as "legal highs", which are chemically distinct analogs of traditionally used drugs not yet legally regulated.

Despite the rapid rise of crypto-markets, internet-based drug trafficking still represents only a small fraction of the world drug trade [5]. Traditional drugs remain the dominant part of international drug trafficking despite the rising popularity of NPS. Market diversification, however, has encouraged polydrug abuse, which poses risks to users as increasing numbers of different and previously unseen substances become available for purchase. Opioid users, for example, are at risk of consuming much stronger and deadlier fentanyl instead of traditional heroin, as synthetic opioids are more economically efficient to produce [4]. For instance, 260 substances classified as NPS were reported in 2012, and that number grew to 483 in 2015, with approximately 80 novel psychoactive substances securing a stable presence on the global drug market. Synthetic drug production is not tied to a particular locale, as these substances do not rely on extracting chemicals from specific plants and can be produced and distributed worldwide from anywhere [4].

Along with the emergence of NPS, "new payment methods" (NPM) have gained traction in legitimate and illegitimate economies alike. NPM include, but are not limited to, prepaid cards, mobile payment services and internet-based payment services which serve as an alternative to traditional payment methods, or offer an innovative way of transferring value between individuals and organizations [6]. For the study presented in this paper, we considered how NPM, specifically mobile payments and crypto-currencies, are used to enable the purchasing of illegal drugs online in a safe and secure fashion. We used our automated web data collection tool, called The Dark Crawler (TDC), to analyze advertisements from illicit drug marketplaces on the public web to assess business practices of this new emerging sector of the illicit drug market. TDC collects webpages from target websites, and then, using customized rules, extracts user-specified content from the pages [7–9]. However, these marketplaces frequently implement strategies to prevent the type of web-analysis we were trying to undertake. Specifically, they utilized Distributed Denial of Service (DDoS) attack protection and captchas. This required modifying our software to be able to satisfy these requirements and allow for (mostly-) automated data-capture. In this study, we used our modified TDC to analyze 28 online drug markets in order to identify types of substances sold, prices, payment methods used and geographic distribution of drug sales.

First, we provide a brief literature review on international drug trafficking and identify conditions which lead to the emergence of markets for new psychoactive substances, and review research on data capture from the internet. Then we outline our data collection methodology, considerations for data capture, followed by an analysis of the data we captured from online drug marketplaces. Finally, the results and limitations are discussed as the paper is concluded.

2 Literature Review

2.1 Drug Prohibition and Drug Market Evolution

The United Nations (UN) played a major role in the internationalization of drug prohibition. The 1961 Single Convention on Narcotic Drugs represents a major milestone of drug regulation which was widely adopted throughout the world [10]. The Single Convention was heavily influenced by the United States due to the American role in the creation of the UN. All 198 nations which signed the Single Convention implemented generally similar drug laws, criminalizing cultivating and refining drugs such as opioids and cannabis. The Single Convention was supplemented by the addition of psychotropic substances (e.g. LSD) to the list of prohibited drugs in the 1971 Convention on Psychotropic Substances [10]. Adherence to the Single Convention differs throughout the world, from lenient drug policies (e.g. Netherlands and Portugal), to stricter hardline approaches (e.g. the UK and Russia) [10–12]. Cannabis, opiates and amphetamine-type stimulants (ATS) are the three most traded and consumed types of drugs today [4].

Worldwide criminalization of traditional drugs has eventually produced a partial market shift to novel designer drugs, also called "legal highs". Producers of such substances can get ahead of legislators and distribute psychoactive substances which are not yet criminalized. Some of the most known novel psychoactive substances are "spice" (synthetic cannabinoids) and "bath salts" (synthetic cathinones) [13–15]. "Spice" refers to an herbal mixture with added synthetic cannabinoids typically marketed as incense, which is sold online and in some countries in specialty stores such as tobacco shops. It first appeared in Europe and the United States in the early 2000s, signaling the rise of the market for "new psychoactive substances" (NPS) [16, 17]. "Spice" is used as an alternative to marijuana, as users reported effects similar to those achieved by smoking cannabis [17]. Along with different drug consumption patterns, emerging NPS markets have resulted in new distribution models. "Spice" and "bath salts" are available for purchase online on specialized websites that can be located through any search engine [15]. The internet has enabled easy access to outlets which sell prescription and non-designer illegal drugs due to regulatory difficulties inherent in controlling businesses which are dispersed among different jurisdictions [17]. Bath salts are typically advertised online as a legal alternative to amphetamine-type stimulant drugs. In the early 2000s, these substances were sold in retail stores owing to their legality at the time, but following their prohibition in 2010s in the US and Europe, the internet has become the primary source of distribution for bath salts and similar new psychoactive substances (NPS) [15, 18, 19]. In the US, synthetic cannabinoids were not widely known before 2009, but by 2010 the interest in "herbal incense" and other names for synthetic cannabinoids skyrocketed, based on internet searches [16].

"Hot spots" of illegal drug trade have emerged on the internet due to the anonymity, global reach, and availability of impersonal payment methods which do not require client identification [20]. Silk Road, the archetypal crypto-market for illicit drugs, was structured similarly to eBay, providing a platform where any vendor can sell their product and any user can make purchases. This business model was made possible due to the ability of clients to make anonymous payments with the crypto-currency Bitcoin, while Silk Road itself was hosted in the encrypted Tor network to maximize anonymity and to minimize risks to participants. The business model pioneered by the Silk Road was revolutionary because it allowed users to access vendor reviews and detailed information about the product, which served to mitigate distrust in an environment of uncertainty [21–25]. Customers of Silk Road have listed a number of reasons for using the marketplace, such as access to a wider range of drugs, convenience, confidence in highly rated sellers and lower prices [21]. Additionally, users found desirable the lack of necessity to turn to street dealers to acquire drugs [26]. User feedback for products they received typically described quality (purity) of the product, delivery speed and packaging stealth [26, 27]. Convenience, high variety and low risk were the key reasons why internet users may purchase recreational drugs online.

While crypto-markets have received extensive attention from journalists, academics and law enforcement alike, there is another form of internet-mediated drug trafficking that has emerged in post-Soviet states, termed "noncontact drug dealing". Like crypto-markets, this method relies on an online storefront where advertisements for drugs are displayed, but the delivery method is different, as well as the fact these marketplaces are hosted openly in the public web and require no registration to access. Instead of shipping, noncontact drug dealing marketplaces utilize pre-arranged drug stashes located throughout the market's area of operation. Once the payment for drugs has been received, marketplace operators transfer instructions on how to locate the stash to clients as an instant message. The seller and the buyer never interact except online, and the buyer makes a money transfer using one of the new digital payment methods with minimal identification measures. This method is discussed in Russian-language literature on online drug trafficking, as well as a Financial Action Task Force (FATF) report, which emphasize the abuse of virtual currencies such as Qiwi, WebMoney and YandexMoney for making these illegal transactions [28–30]. The ability of these marketplaces to operate openly is possibly explained by bureaucratic and legislative obstacles encountered by Russian law enforcement in policing online drug dealing [31–33].

2.2 Noncontact Drug Dealing

The "noncontact" method of drug dealing has gained prominence due to the ability of individuals to use "new payment methods" with minimal identification measures. By communicating with drug vendors over the internet through discussion forums, online storefronts and marketplaces, drug users are able to procure prohibited substances without ever meeting the seller. The buyer and the seller agree on a place, which is used as a hiding spot for drugs, and a time for pickup. Once the buyer has made the payment, they receive instructions on where to find the stashed drugs [34]. Noncontact drug dealing appears to be predominantly used in Russia and neighboring Commonwealth of Independent States (CIS) countries. The Financial Action Task Force (FATF) report which included a description of the noncontact drug dealing method used an example that referred to Russia and Tajikistan [34]. Noncontact drug dealing has been described extensively in Russian academic journals. Drug market actors have started using mobile payment providers, such as Qiwi, WebMoney or YandexMoney, which enable money transfers between individuals [29]. A source in Russian law enforcement claimed that hand-to-hand drug dealing has become virtually non-existent and drug traffickers have largely moved on to noncontact drug dealing [35].

The essence of the noncontact drug dealing method can be described as follows: (1) the buyer makes an order online (through a website or an instant messenger);

(2) the buyer then makes a payment with a virtual currency or as a mobile money transfer; and (3) the buyer receives instructions as an SMS or an instant message on where to find the stash with drugs [36]. This method of drug distribution became commonplace because it significantly reduces the risk of becoming the target of buy-and-bust operations [30]. The current study analyzed 28 online drug marketplaces all of which used the noncontact drug dealing method. The data pertaining to geographic distribution of online marketplaces, drug types, prices, amounts and payment methods were extracted to evaluate criminal risks posed by the noncontact method of drug dealing coupled with the ability to make anonymous payments.

2.3   Data Capture from Web

The information age brought about persistent connectivity, and with it came the ever-present digital cataloging of our activities—whether legitimate or illegal. Taking advantage of these "digital traces" allows researchers to study online activities of criminals in their natural environment—so-called "convergence settings", rather than relying on potentially unreliable self-report data in surveys or interviews from a subset of "unsuccessful" criminals who were apprehended [5, 37]. Manual data collection is a labor-intensive process as it may require researchers to analyze hundreds and thousands of webpages which need to be saved locally, coded, and finally subjected to analysis. Despite this, manual data collection can provide large and rich datasets, which proved valuable for an in-depth analysis of online criminal activities [5, 37]. For example, Buskirk and colleagues sampled a crypto-market called Agora, at the time the largest "dark net" market, by manually saving webpages and extracting the data with a Microsoft Excel VBA macro [38]. This method allowed authors to automatically categorize nearly 80% of listings, which shows that manual data collection is not necessarily an obstacle to efficient internet-based research [38].

Automated data collection relies on software that creates local copies of the target content hosted online, a process called "mirroring", and extracting the relevant data from saved webpage content in a process called "scraping" [39]. Also known as web crawling, this process has typically been focused on online social networks such as blogs and forums, and also websites, often for consumer research and marketing purposes. In a review of research on web crawling most of studies were published in early 2000s and were concerned with conventional content such as cooking recipes and movies in the surface web [40]. The context of these studies is significantly different from criminal content hosted on the public web or the "dark web", where site administrators may take active steps to protect the website contents from access and indexing by automated means. The original Silk Road used cookies that allowed users to stay logged in for a week before expiring, which enabled researchers to bypass captchas while crawling the website [41]. Unlike the first iteration of Silk Road, Silk Road 2 did not use a captcha or require manipulating website cookies to perform data capture, enabling researchers to produce what the authors claims is a complete crawl of the website [23]. This approach however was criticized due to the instability of websites hosted in the Tor network undermining automated data collection methods which possibly result in incomplete datasets [38].

Web crawling, or "mirroring", is typically performed in the following fashion: starting with a single webpage, the program indexes all hyperlinks within it then follows those links, repeating the process for each subsequent webpage within limits set by the user [39]. Ready-made solutions capable of mirroring webpages exist, such as HTTrack [23]. However, HTTrack and similar software will only mirror webpages, without analysis, thereby requiring subsequent steps for data cleaning and conversion of the unstructured data into structured data. Scraping can also be performed by separate programs, called scrapers, which can parse webpages to identify relevant content, such as usernames attached to forum posts or drug names [39]. Custom-written web crawlers tend to have more functionality than freely available solutions, such as the crawler DATACRYPTO which was specifically created for studying drug listings on the original Silk Road. This crawler both mirrored and scraped the webpages, building a database of drug listings, vendor information and buyer feedback. Creating a custom web crawler from scratch requires significant investments and may not be feasible for all researchers [5, 39].

For this study, the contents of target websites were downloaded using a custom-written web crawler and scraper called The Dark Crawler (TDC). TDC has previously been used to study hacker forums and extremist websites and to identify terrorist and child exploitation content on the dark web [9, 42–44]. In addition to automated data collection, TDC allows integrating other analytical tools for studying the extracted content. For example, natural language processing (NLP) supplemented with parts of speech (POS) tagging were used to calculate sentiment scores for posts on hacking forums that reference attacks against critical infrastructure such as government facilities, banks and hospitals [42]. Sentiment analysis showed discussions of prominent topics, such as financially motivated attacks against banks, or exploitation of vulnerabilities in government systems, corresponded to actual data on these incidents—what forum members discuss are types of attacks that are typically carried out [42]. Understanding the threat landscape of such attacks can help potential targets prepare for them and mitigate the damage. Another advantage of using an automated web crawler is the lack of necessity to manually review potentially disturbing media to establish their criminal content. For instance, identifying child exploitation and terrorism-related images is possible through image hashing, where each image is assigned a hash value that can be compared against law enforcement databases of known illegal images [9].

2.4 Studying Online Drug Trafficking

Being the most notorious drug marketplace, the Silk Road attracted much attention from academics. This drug marketplace was one of the first to be studied in-depth before its shutdown by the authorities. The number of methods and approaches used was quite varied, from qualitative analyses of attitudes towards online drug purchases, to analyzing trends among drug listings by saving snapshots of the website for a given period [17, 38]. By collecting vendor names and PGP keys (cryptographic sequences used to authenticate a user) through an automated web crawler, researchers were able to describe organizational structure of online drug distribution networks [22]. In order to evaluate the impact of Silk Road's closure on the "darknet" drug trafficking economy, the number of vendors on competitor websites was analyzed through approximately one month period. Despite the shutdown of a major distribution hub, the drug market actors were able to quickly adapt and to relocate to competing markets, as demonstrated by an explosive growth of the number of vendors on them [38]. Multiple studies categorized the illicit goods offered on these marketplaces, ranging from drugs to pornography, stolen data and weapons [23, 24]. We chose to focus on extracting and analyzing sales data – specifically drug names, prices, quantities and regions where they were available. This methodology allows to present an overview of the noncontact drug dealing problem and its extent, as well as to consider implications of this method for combatting drug trafficking internationally.

Furthermore, as stated previously, noncontact drug dealing is a topic of extensive discussion among Russian-speaking academics and law enforcement profession-als [28–30, 35, 36, 45–48]. However, the studies are primarily concerned with organizational aspects of noncontact drug dealing [28, 29, 36, 46–48], as well as investigative methods and practices [29, 46, 48]. Few studies attempt to characterize the drug market facilitated through these online marketplaces in terms of drug types, prices and volumes beyond general trends (e.g. a 9.5% increase in consumption of synthetic drugs between 2013 and 2015) [35, 36, 45]. By leveraging data collection capabilities of the web crawler, we were able to provide an overview of the market based on the publicly available drug listings data [23].

3 Methods

As this is an exploratory study, the intention was to identify market prices, the scale of the problem (i.e. how widespread non-contact drug dealing is), and which payment methods were used to pay for the drugs. Specifically, the goal is to systematically go through entire online websites and extract the required content so the drug prices can be analyzed. To do this, 28 websites were selected (Sect. 3.1). However, before data could be captured via our existing The Dark Crawler (TDC) infrastructure, several challenges had to be overcome as the target websites had implemented techniques to specifically prevent automated bots from entering the site. After these changes were implemented (Sect. 3.2) we were able to capture the required pages (Sect. 3.3), and defined rules (Sect. 3.4) which allowed us to extract the drug prices, locations and volumes into an analyzable dataset.

3.1   Data Selection

Exploratory research led us to public Russian drug forums as a starting point, where advertisements of drug markets in our sample were discovered. A clear pattern emerged among the websites found—their designers relied on largely the same webpage template with minor variations such as background images. Additionally, these websites were named similarly and hosted on the same top-level domain *.biz* (for example, *narco24.biz, kumar24.biz*, etc.). Although multiple domains were used, *.biz* appeared to be the most prevalent—searches for "24 biz" proved to be the most fruitful, while searches for "*24 pw*" or "*24 cc*" yielded less relevant results. Overall, over 47 open drug market websites were identified by searching "24 biz" from 100 search hits spread out on 10 pages of the results from the Russian search engine Yandex. Inclusion criteria were established, where an open drug market website had to (1) be public (accessible without registration) and (2) display drug advertisements. Between June 2016 and August 2016, only 28 out of 47 (59.5%) drug market websites remained available long enough to be captured, whereas 19 (40.5%) had to be omitted as no advertisements were displayed.

3.2   Data Collection Challenges

TDC has been used to crawl webpages and extract structured data from them (for examples, see [9, 42–44]), however the websites that were picked for this study differed from previously studied websites in that these were protected by DDoS browser checks.

There are multiple commercial solutions available to defend against DDoS attacks, for example, by shielding the target website with proxy servers that distribute the incoming traffic between themselves and balance the load, or requiring the browser to solve JavaScript calculations before the page is displayed. If this ability to solve JavaScript calculations is not present, then it is highly likely that the requestor is an automated web crawler. The online marketplaces for illegal drugs sampled for the current study utilized DDoS protection offered by a major commercial provider also relied on by legitimate businesses. By integrating the ability to bypass DDoS browser checks into the TDC, we have significantly expanded the pool of websites the data can be collected from. TDC initially was relying on simple GET requests executed in parallel via multiple threads to retrieve multiple HTML pages simultaneously (Fig. 1). However, while this method does retrieve the HTML served by the server, it does not interpret (i.e. execute) any of the content on it. Any JavaScript code that is required does not run (because it is not even retrieved, unless it is embedded into the HTML), thus the checks that run against the browser would fail. To get around this problem, the GET method was replaced by the customizable web-browser engine Chromium. See Fig. 2 for a high-level overview of the structure of TDC.

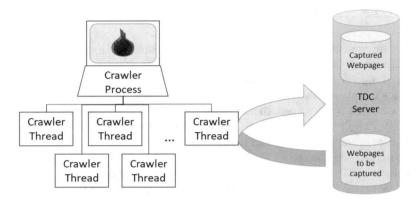

Fig. 1 The Dark Crawler high-level overview (original)

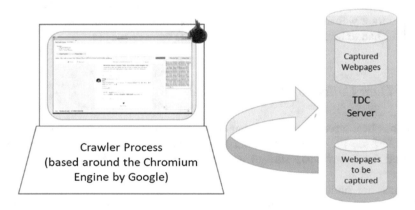

Fig. 2 The Dark Crawler high-level overview (modified)

To capture the data from these websites, the webpage-retrieval engine used by TDC was replaced with an actual browser capable of being automated. In this fashion, the automated browser, called VisualChromium, could access and render the webpages just like any other browser would, in the process solving any JavaScript required. When the browser-checks were done, and the webpage loaded, the automation kicked in and the HTML retrieved was analyzed. Some pages required the solving of captchas to confirm the user is human and the retrieval is not an automated request. As pages are being retrieved, TDC checks if each page meets certain conditions, such as specific text (not) being present. If these conditions are met, TDC moves onto the next page, and if not, the data capture stops and awaits user action. A condition which was present on "normal" pages but not captcha pages allowed the Crawler to recognize captcha pages and to wait for user action before

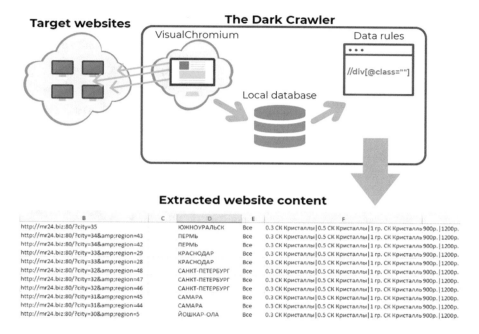

Fig. 3 A summary of the data capture process

continuing. In the future, automated captcha solving will be investigated, but for the sites downloaded in this study, the websites did not use captchas, and thus this was not an issue.

After the "mirroring" process was complete, the captured data was extracted based on user-specified data rules (see Fig. 3 for an overview, and Sect. 3.4 for more details). All rules were applied to all data, producing a .csv format file compatible with Microsoft Excel. Each column represents a data element (e.g. drug names) and rows represent single web-pages. Particularities of data collection and using the Dark Crawler are discussed below.

3.3 Final Data Collection

The Dark Crawler is an automated data collection tool which is able to download entire contents of webpages utilizing user-specified rules. Given a list of webpages, the Crawler will traverse them and parse them apart. The process is as follows:

```
<HTML>                                  <HTML>
 <DIV>                                   <DIV>
  <A href="index.html">Home</A>           <A href="index.html">Home</A>
</HTML>                                  </DIV>
                                        </HTML>
```

a) HTML tags are not closed properly in a lot of websites   *b) Standard HTML*

Fig. 4 Real-life HTML vs the standard. (**a**) HTML tags are not closed properly in a lot of websites. (**b**) Standard HTML

- For each webpage within the Queue, repeat until the Queue is empty

  - the Crawler gets a webpage from the queue of webpages to retrieve;
  - it retrieves the webpage, resulting in an HTML string;
  - the webpage HTML is cleaned up, with invalid (or unclosed) tags fixed (see Fig. 4 for an example);
  - the HTML is parsed according to data capture rules and relevant information is identified, such as links or images
  - each link within the HTML is added to the queue of future pages to be retrieved

- Rules are applied to the cleaned HTML to extract user-defined pieces of data (in this case, drug names, prices, amounts, etc.) into a spreadsheet (see Sect. 3.4 for details).

As the captured data is stored in a structured database, the data extraction process can be repeated after the data capture process, and the data extracted in multiple ways, e.g. by searching for specific keywords or extracting certain types of data elements (e.g. drug names). The resultant information is then saved in the form of a matrix, with each webpage in a row, and each extracted data-element in a column. This data can then be subjected for further analysis by, for example, geo-mapping the distribution of drug sale locations.

3.4   Data Extraction

Landing pages of drug marketplaces in our sample were structured in a similar way. Each drug advertisement was placed inside a container that showed drug name, price, quantity, whether the goods were in stock, and a "Buy" button which would take the user to a page with transaction details. An example of such a webpage is shown in Fig. 5. Each webpage element can be targeted for capture by defining data rules. Some of the websites in the sample modified the default webpage template (e.g. by restructuring and renaming elements of the webpage), however the overall structure was consistent across all 28 sites studied. TDC downloaded each page, cleaning up unnecessary data, resulting in a structured table containing data from these webpage elements. This enables the user to quickly extract all the required information from a webpage, significantly reducing the amount of manual work required to preserve these data.

The Dark Crawler identifies relevant information within the webpage by relying on data rules specified by the user. A data rule is a {*path, pattern*} combination, which uses the XPaths standard for *path* querying trees and the XQuery standard for the *pattern*s, see Sects. 3.4.1 and 3.4.2 respectively for more details. This process allows selecting elements from a webpage to be downloaded and stored in a database.

Using Paths to Select Webpage Elements

A hierarchy of branches from the root makes up a path, which follows the *XPaths* standard, where a path results in zero or more nodes in a tree. To identify the drug name which is priced at "900" in Fig. 5, the Crawler identifies distinct containers on the webpage, traversing series of branches (*div, ul, li,* and *div*), where the path would be "*/html/body/div/section/ul/li/div*". Figure 6 shows how this process identifies 4 drug names located on this webpage, one per container. To select a specific drug name from the list, the path is modified to target one of the <*div*> tags under <*li*>. The resultant rule then would be "*/html/body/div/section/ul/li/div[2]*" which directs the Crawler to select the second *div* tag from the branch. By iterating through all the *div* tags in this fashion, all prices can be extracted for all the drugs on the webpage. A similar strategy is used to extract all other required content.

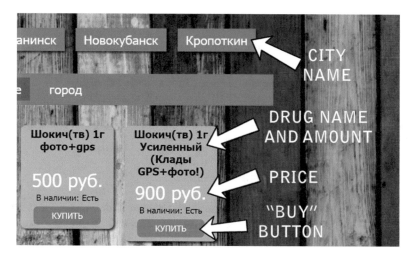

Fig. 5 A typical drug marketplace landing page with advertisements displayed

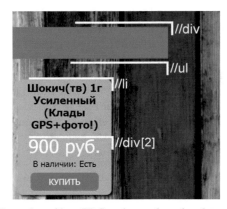

a) An illustration of how TDC navigates through webpage elements

```
<!DOCTYPE html>
<html lang="ru">
▶ <head>…</head>
▼ <body>
  ▶ <section id="userpanel">…</section>
  ▶ <section id="logo">…</section>
  ▼ <div class="content">
    ▶ <section id="menu">…</section>
    ▶ <section id="mobile">…</section>
    ▶ <section id="index">…</section>
    ▼ <section id="index">
      ▼ <ul id="products">
        ▶ <li>…</li>
        ▼ <li>
            <div class="name">
                        3таб. EXTASY XTC (Европа)</div> == $0
          ▶ <div class="price">…</div>
            <div class="size">B
                        наличии: Есть                  </div>
          ▶ <div class="buy btn">…</div>
          </li>
        ▶ <li>…</li>
```

b) The HTML code of a webpage, shown as a tree-like structure

Fig. 6 HTML content as a tree on a similar drug market website. (**a**) An illustration of how TDC navigates through webpage elements. (**b**) The HTML code of a webpage, shown as a tree-like structure

Applying Patterns to Filter Results of Paths

Targeting the webpage element allows the Dark Crawler to download the desirable content, but not necessarily in the proper format. The *rule* selects the necessary element, but cannot select any specific parts of it. This is problematic, for example, where the price would be extracted as "900 руб.", and not just "900". To remove unnecessary text or to perform calculations with the result of operations with the *path* rule, *patterns* are used, which in turn follow the standard language of XQuery. For an example, working with Fig. 5, the currency is specified after the numeric value, which is in rubles for all the advertisements in the sample. Moving down the HTML branches to */li/div class="price"* would produce the result "900 руб.", where only "900" is necessary. Specifying a pattern in addition to the path will prune any unnecessary data. The pattern "(?<RESULT1>.*) руб." is created to remove the text following the numeric value.

4 Results

4.1 eDrugs

Once the 28 websites were captured, and rules set up, 935 drug advertisements were extracted. Most of the sample (N = 839, 89.7%) was represented by small quantities intended for personal use ranging between 0.3 and 10 g per order. However, a fraction of the sample (N = 97, 10.3%) contained larger quantities of drugs, advertisements for which were marked as "wholesale" or "pre-order". Drug amounts in "wholesale" orders ranged between 3 and 1000 g, where smaller marketplaces typically sold a few grams of substances per order, and larger marketplaces offered amounts upwards of tens and hundreds of grams per posting. Each website displayed between 1 and 250 drug postings, with 29 on average. Among types of substances being sold, wholesale orders and consumer-sized orders taken together, amphetamine-type stimulants made up 58.5% (n = 547), synthetic cannabinoids—22.4% (n = 210) and natural cannabinoids, i.e. marijuana products—16.3% (n = 153). The rest of the sample, 2.9% (n = 25) was represented by hallucinogens, advertised by a single marketplace as "acid". This is shown in Figs. 7 and 8. Regional differences between drugs offered for sale are visible in Fig. 8. More variation in substance types can be observed in the Urals region, and natural cannabinoids appear more frequently.

4.2 Markets

Three categories of drug marketplaces become immediately evident – large (sales revenue over $4000), medium ($800–$1600) and small (under $800). The distribution of total drug prices across marketplaces and average drug prices across the Central region respectively are shown in Figs. 9 and 10. While large marketplaces conducted business in multiple cities and different geographical areas, medium marketplaces were limited to a more meager area of operation, and small marketplaces focused exclusively on a single city and/or small suburban towns located closest to it. Drug prices across the country remained surprisingly consistent, which could be explained by established equilibrium prices or price fixing.

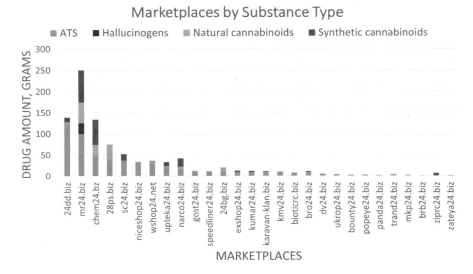

Fig. 7 Marketplaces by substance type

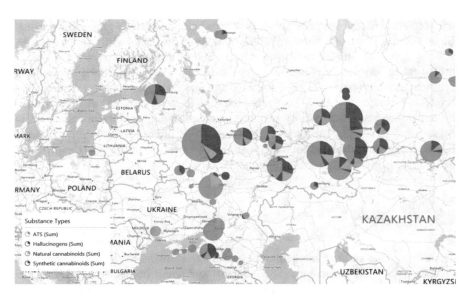

Fig. 8 Substance types spread across the region

Fig. 9 Marketplaces by total price

Fig. 10 Average prices across the region

4.3   Payment Methods

Noncontact drug dealing literature implicates virtual currencies such as Qiwi, WebMoney, YandexMoney and Bitcoin in internet-mediated drug trafficking, as these virtual currencies offer convenient ways to pay without being subjected to the same standards for identification as bank payments would be. However, in our sample of open web drug markets only Qiwi and Bitcoin were used. Furthermore, Qiwi, a mainstream virtual currency marketed as a vehicle for personal money transfers and utility bill payments, was used more often—on 28 marketplaces, as opposed to 20 for Bitcoin (see Figs. 11 and 12). This is perhaps due to the fact Qiwi is much more "user-friendly", where anyone can create an online wallet without any identification. Anonymous Qiwi wallets are limited to $250 per transaction up to $665 per month, and personal money transfers (person-to-person transactions) are prohibited by federal legislation. Nevertheless, these personal money transfers play a key role in facilitating online drug trafficking, as open web drug markets' business model relies on simple and straightforward payments available to any internet user without a level of technical and financial competency required to use crypto-currencies. Even without taking into account the illegal nature of goods being sold on these marketplaces, payments for these goods themselves are against the law. A number of procedural and legislative obstacles exists which prevent efficient policing of the internet environment by the Russian law enforcement, as mentioned above.

Fig. 11 Marketplaces by payment methods

Fig. 12 Payment methods across the region

5 Discussions and Conclusion

The digital economy has reshaped trade in both conventional and illicit goods. International drug prohibition regimes criminalizing traditional drugs throughout most of the world forced drug markets to adapt and to innovate, resulting in the rise of novel synthetic drugs acting as analogs of traditional substances capable of being produced anywhere. Although the majority of world drug trafficking is still made up of traditional drugs, the rising popularity of novel psychoactive substances and innovative methods of their distribution emerging online represent a new worrisome trend. Taking advantage of opportunities afforded by the digital economy and e-commerce, drug traffickers have capitalized on the ability to sell novel, not yet regulated substances with the use of new payment technologies that enable their clients to pay with relative anonymity. We analyzed 28 online drug marketplaces which sold both traditional and novel psychoactive substances. While amphetamine-type stimulants and synthetic cannabinoids dominated the sample, traditional marijuana products and some hallucinogens were also advertised. These marketplaces primarily targeted the average consumer with drug amounts intended for personal use, however larger marketplaces also offered business-sized amounts of drugs for purchase.

Drug marketplaces which utilize noncontact drug dealing advertise and distribute drug products through internet sites, similarly to crypto-markets, but rely on a different delivery method, i.e. pre-arranged stashes. The ability of marketplace clients to carry out anonymous transactions via Qiwi and Bitcoin enables this business model to exist. Relative to crypto-markets, these websites are even easier to access, and the ability to pay with a common virtual currency enables anyone with a small amount of cash and a mobile phone to make a purchase. Operators of drug marketplaces in our sample took active steps to prevent automated access to their websites, which were circumvented by designing a separate application acting as web browser that can be used to solve the captcha and pass a DDoS protection browser check so that the crawl may continue. Automated data collection enabled us to download and analyze 935 advertisements for drugs, identifying contours and practices of the market. Although some of the websites identified as online drug marketplaces became unavailable during the data collection period, the information from the remainder of the sample was sufficient to chart and map the extent of the problem. Online drug marketplaces appear to operate throughout major cities and population clusters, where a higher variety of drugs is offered. Expanding the sample to include more drug marketplaces and/or locations would result in a more complete picture of the market. Such studies can be used to identify innovations in drug distribution, as well as to gauge consumption patterns among drug users.

The total number of noncontact drug-dealing marketplaces is unknown, therefore representativeness of our sample cannot be stated with certainty (935 postings across 28 marketplaces). Nearly half of the drug marketplaces initially identified for data capture became unavailable or had no postings displayed during the data collection period, pointing to instability of these websites perhaps due to market forces, business practices such as exit scams or law enforcement pressure. While these results may not paint a comprehensive picture, this study offers a glimpse into the noncontact drug dealing market which exists due to the ability of marketplaces operators to exploit virtual currencies with low (or none, in case of Bitcoin) customer identification standards. Since this study only considered Russian-language online drug marketplaces, external validity of these results may be low due to differences in legislation, availability of payment providers or drug trafficking routes. Nevertheless, the delivery method, i.e. pre-arranged stashes, represents an innovation in drug distribution that drug traffickers in other countries may also utilize assuming there are easy payment methods in place for internet users to transfer value (digital currency in most cases) from one person to another.

From a law enforcement perspective this type of non-contact drug purchase poses several challenges in identifying the dealer. While the delivery method would prevent the de-anonymization of the seller, the payment method, both Qiwi and bitcoin, could be traced. Although the money could be traced when the buyer transfers Qiwi to the dealer, as both the sender and recipient use mobile phones, the dealer can add a layer of security to the transaction by immediately converting Qiwi to physical currency through ATM withdrawals, and through the use of burner-phones. This would provide only a small window of opportunity for law enforcement to identify the identity and location of the dealer, before the money is withdrawn and the burner phone disposed. In the event the buyer pays with bitcoins, the identification of the dealer becomes much more challenging, as it takes significant effort to trace Bitcoin payments through tumblers, and eventually to a Bitcoin exchange, where the money might not be withdrawn for years. Finally, law enforcement could try to identify and to apprehend the author/owner of the website which peddles NPS, although, given the large number of cities where NPS is available through online non-contact methods, it is very likely that the owner of the website is acting as a middle-man between the buyer and the actual on-site dealer. Like crypto-markets, larger drug marketplaces in our study also offered business-to-business purchases with drugs intended for resale, emphasizing the scalability of this business model which can serve both individuals and subsequent drug traffickers in the chain of distribution.

References

1. P. Mullan, Who uses digital currency? in *The Digital Currency Challenge: Shaping Online Payment Systems Through US Financial Regulations* (Palgrave Pivot, New York, 2014), pp. 13–15
2. T. Tropina, Fighting money laundering in the age of online banking, virtual currencies and internet gambling. ERA Forum **15**(1), 69–84 (2014)
3. D. Décary-Hétu, L. Giommoni, Do police crackdowns disrupt drug cryptomarkets? A longitudinal analysis of the effects of Operation Onymous. Crime Law Soc. Chang. **67**(1), 55–75 (2017)
4. United Nations Office on Drugs and Crime, World Drug Report 2017 (2017), https://www.unodc.org/wdr2017/index.html
5. J. Aldridge, D. Décary-Hétu, Hidden wholesale: the drug diffusing capacity of online drug cryptomarkets. Int. J. Drug Policy **35**(C), 7–15 (2016)
6. Financial Action Task Force, *Guidance for a Risk-Based Approach to Prepaid Cards, Mobile Payments and Internet-Based Payment Services* (FATF/GAFI, Paris, 2013), http://www.fatf-gafi.org/publications/fatfrecommendations/documents/rba-npps-2013.html
7. B. Monk, J. Mitchell, R. Frank, G. Davies, Uncovering Tor: an examination of the network structure. Secur. Meas. Cyber Netw. **2018**, 4231326 (2018)
8. B. Westlake, M. Bouchard, R. Frank, Assessing the validity of automated webcrawlers as data collection tools to investigate online child sexual exploitation. Sex. Abus. **29**, 685 (2015)
9. A.T. Zulkarnine, R. Frank, B. Monk, J. Mitchell, G. Davies, Surfacing collaborated networks in dark web to find illicit and criminal content, in *Intelligence and Security Informatics (ISI) Conference*, Arizona (2016)
10. F. Mena, D. Hobbs, Narcophobia: drugs prohibition and the generation of human rights abuses. Trends Organised Crime **13**(1), 60–74 (2010). https://doi.org/10.1007/s12117-009-9087-8
11. D.R. Bewley-Taylor, The American crusade: the internationalization of drug prohibition. Addict. Res. Theory **11**(2), 71–81 (2003). https://doi.org/10.1080/1606635021000021377
12. E. Crick, Drugs as an existential threat: an analysis of the international securitization of drugs. Int. J. Drug Policy **23**(5), 407–414 (2012). https://doi.org/10.1016/j.drugpo.2012.03.004
13. J. Buchanan, Ending drug prohibition with a hangover? Br. J. Community Justice **13**(1), 55–74 (2015)
14. D. Perrone, R.D. Helgesen, R.G. Fischer, United States drug prohibition and legal highs: how drug testing may lead cannabis users to spice. Drugs: Educ., Prev. Policy **20**(3), 216–224 (2013). https://doi.org/10.3109/09687637.2012.749392
15. K. Meyers, Ö. Kaynak, E. Bresani, B. Curtis, A. McNamara, K. Brownfield, K.C. Kirby, The availability and depiction of synthetic cathinones (bath salts) on the Internet: do online suppliers employ features to maximize purchases? Int. J. Drug Policy **26**(7), 670–674 (2015). https://doi.org/10.1016/j.drugpo.2015.01.012.GLOBAL CRIME 25

16. B. Curtis, K. Alanis-Hirsch, Ö. Kaynak, J. Cacciola, K. Meyers, A.T. McLellan, Using Web searches to track interest in synthetic cannabinoids (aka 'herbal incense'). Drug Alcohol Rev. **34**(1), 105–108 (2015). https://doi.org/10.1111/dar.12189
17. P. Griffiths, R. Sedefov, A. Gallegos, D. Lopez, How globalization and market innovation challenge how we think about and respond to drug use: 'Spice' a case study. Addiction **105**, 951–953 (2010). https://doi.org/10.1111/j.1360-0443.2009.02874.x
18. J. Gershman, A. Fass, Synthetic cathinones ('bath salts'): legal and health care challenges. P T **37**(10), 571–595 (2012)
19. L. Karila, B. Megarbane, O. Cottencin, M. Lejoyeux, Synthetic cathinones: a new public health problem. Curr. Neuropharmacol **13**(1), 12–20 (2015)
20. A. Lavorgna, Internet-mediated drug trafficking: towards a better understanding of new criminal dynamics. Trends Organised Crime **17**(4), 250–270 (2014). https://doi.org/10.1007/s12117-014-9226-8
21. M.J. Barratt, J.A. Ferris, A.R. Winstock, Use of Silk Road, the online drug marketplace, in the United Kingdom, Australia and the United States. Addiction **109**(5), 774–783 (2014). https://doi.org/10.1111/add.12470
22. J. Broséus, D. Rhumorbarbe, C. Mireault, V. Ouellette, F. Crispino, D. Décary-Hétu, Studying illicit drug trafficking on Darknet markets: structure and organisation from a Canadian perspective. Forensic Sci. Int. **264**, 7–14 (2016). https://doi.org/10.1016/j.forsciint.2016.02.045
23. D.S. Dolliver, Evaluating drug trafficking on the Tor Network: Silk Road 2, the sequel. Int. J. Drug Policy **26**(11), 1113–1123 (2015). https://doi.org/10.1016/j. drugpo.2015.01.008
24. A. Phelps, A. Watt, I shop online – recreationally! Internet anonymity and Silk Road enabling drug use in Australia. Digit. Investig. **11**(4), 261–272 (2014). https://doi.org/10.1016/j.diin.2014.08.001
25. M.V. Hout, T. Bingham, 'Surfing the Silk Road': a study of users' experiences. Int. J. Drug Policy **24**(6), 524–529 (2013). https://doi.org/10.1016/j.drugpo.2013.08.011
26. M.C. Van Hout, T. Bingham, Responsible vendors, intelligent consumers: Silk Road, the online revolution in drug trading. Int. J. Drug Policy **25**(2), 183–189 (2014). https://doi.org/10.1016/j.drugpo.2013.10.009
27. J. Van Buskirk, A. Roxburgh, M. Farrell, L. Burns, The closure of the Silk Road: what has this meant for online drug trading? Addiction **109**, 517–518 (2014). https://doi.org/10.1111/add.12422
28. A.L. Osipenko, P.V. Minenko, Investigative counteraction to illegal drug trafficking by telecommunication devices (in Russian). Bull. Voronezh Institute of MVD **1**, 151–155 (2014)
29. A.V. Puptseva, Problematic issues of detecting crimes in the domain of illegal drug trafficking with the use of wireless communication devices (in Russian), in *VIII International Scientific and Practice Conference*, Tyumen, 15 Feb 2016
30. A.V. Ryasov, The issues of illegal sale of drugs using information and telecommunication networks (in Russian). Vestnik SevKavGTI **16**, 197–199 (2014)
31. A.N. Kolycheva, Certain aspects of preserving evidentiary information stored on Internet resources (in Russian). Bull. Udmurt University. Economics Law **2**(27), 109–113 (2017)
32. L.M. Kryzhanovskaya, Interaction investigators and officers inquest in the production of selected investigative actions on cases of illegal trafficking in narcotic drugs (in Russian). Theory Pract. Soc. Dev. **1**, 1–4 (2008)
33. F.P. Vasilyev, Modern features of interpretation of law enforcement interaction in Russia and the need for improvement (in Russian). Innov. Sci **3**(2), 102–110 (2017)
34. Financial Action Task Force, *Financial Flows Linked to the Production and Trafficking of Afghan Opiates* (FATF/GAFI, Paris, 2014), http://www.fatf-gafi.org/documents/news/financial-flows-afghan-opiates.html
35. I.A. Sementsova, A. I. Fomenko, Internet environment as a method of trafficking in illegal drugs, psychoactive substances or their analogs (in Russian), in *VIII International Scientific and Practice Conference*, Tyumen, 10 Feb 2016
36. D.A. Donika, Noncontact method of drug distribution (in Russian). Int. J. Exp. Educ. **6**, 17–18 (2014), http://cyberleninka.ru/article/n/sbyt-narkoticheskih-sredstv-beskontaktnym-sposobom

37. M. Barratt, J. Aldridge, Everything you always wanted to know about drug cryptomarkets (but were afraid to ask). Int. J. Drug Policy **35**, 1–6 (2016)
38. J. Van Buskirk, S. Naicker, A. Roxburgh, R. Bruno, L. Burns, Who sells what? Country specific differences in substance availability on the Agora cryptomarket. Int. J. Drug Policy **35**, 16–23 (2016)
39. J. Aldridge, D. Decary-Hetu, Sifting through the net: monitoring of online offenders by researchers. Eur. Rev. Organised Crime **2**(2), 122–141 (2015)
40. H. Chen, *Dark Web*, Integrated Series in Information Systems, vol 30 (Springer, New York, 2012)
41. N. Christin, Traveling the Silk Road: a measurement analysis of a large anonymous online marketplace (2012). https://doi.org/10.21236/ada579383
42. M. Macdonald, R. Frank, J. Mei, B. Monk, Identifying digital threats in a hacker web forum, in *2015 IEEE/ACM International Conference on Advances in Social Networks Analysis and Mining (ASONAM)* (2015), pp. 926–933
43. M. Wong, R. Frank, R. Allsup, The supremacy of online white supremacists – an analysis of online discussions by white supremacists. Inf. Commun. Technol. Law **24**, 41–73 (2015)
44. K. Joffres, *Disruption Strategies for Online Child Pornography Networks* (Library and Archives Canada, Ottawa, 2012)
45. A.I. Anapolskaya, Characterizing typical methods and traces of crimes related to illegal drug trade (in Russian). Eur. Union Sci.: Jurisprud. **8**(17), 136–138 (2015), http://cyberleninka.ru/article/n/harakteristika-tipichnyh-sposobov-i-sledov-soversheniya-prestupleniy-svyazannyh-s-nezakonnym-oborotom-narkotikov
46. M.V. Kondratiev, V.K. Znikin, Operational investigative characteristics of crimes related to the illegal sale of drugs (in Russian). Bull. Kemerovo State University **1**(61), 248–255 (2015), http://cyberleninka.ru/article/n/operativno-rozysknaya-harakteristika-prestupleniy-svyazannyh-s-nezakonnym-sbytom-narkoticheskih-sredstv
47. O.N. Korchagin, D.K. Chirkov, A.C. Litvinenko, Synthetic drugs in Russia as a threat to national security (in Russian). Actual Probl. Econ. Law **1**(33), 245–253 (2015), http://cyberleninka.ru/article/n/sinteticheskie-narkotiki-v-rossii-kak-realnaya-ugroza-natsionalnoy-bezopasnosti
48. A.V. Shebalin, Specifics of conducting a preliminary inquiry into illegal sales of drugs via noncontact methods (in Russian). Curr. Issues Fighting Crim. Other Offenses. **1**, https://xn%2D%2D90ao9d.xn%2D%2Db1aew.xn%2D%2Dp1ai/science/Izdanija_BJUI_MVD_Rossii/Materiali_nauchno_prakticheskih_konferen/AKTUALNIE_PROBLEMI_BORBI_S_PRESTUPLENIJA

Inferring Systemic Nets with Applications to Islamist Forums

David B. Skillicorn and N. Alsadhan

Abstract Open-source intelligence often requires extracting content from documents, for example intent and timing. However, more interesting and subtle properties can be extracted by directing attention to the thought patterns and framing that is implicitly present in the writings of groups and individuals. Bag-of-words representation of documents are useful for information retrieval, but they are weak from the perspective of intelligence analysis. We suggest that systemic functional linguistics, with its focus on the purpose an author intends for a document, and its abstraction in terms of choices, is a better foundation for intelligence analysis. It has been limited in practice because of the difficulty of constructing the systemic nets that are its representation of these choices. We show that systemic nets can be constructed inductively from corpora using non-negative matrix factorisation, and then apply this to infer systemic nets for language use in islamist magazines published by three different groups: Al Qaeda, Daish (ISIS), and the Taliban. We show that the structures captured are also present in posts in two large online forums: Turn to Islam and Islamic Awakening, suggesting a widely held mindset in the Islamic world.

1 Motivation

Applying intelligence collection and analysis strategies to open source data is an obvious strategy because of the availability of vast numbers of documents online: web pages, but also social network status updates, tweets, forum posts, blogs, and podcasts. Leveraging this data requires solving two problems: (1) finding documents relevant to a subject of interest, and (2) extracting the intelligence content implicit in those documents.

D. B. Skillicorn (✉) · N. Alsadhan
School of Computing, Queen's University, Kingston, ON, Canada
e-mail: skill@cs.queensu.ca

© Springer Nature Switzerland AG 2020
M. A. Tayebi et al. (eds.), *Open Source Intelligence and Cyber Crime*, Lecture Notes in Social Networks, https://doi.org/10.1007/978-3-030-41251-7_5

The first problem, information retrieval, has been solved using the bag-of-words approach to representing the content of each document, followed by large-scale index search and careful ranking of the document set. Web search businesses depend on this as a crucial technology on which they build monetized services such as focused advertisements.

The bag-of-words representation of text has proven extremely successful, even for languages such as English where word order is crucial to meaning. However, it is less useful for extracting intelligence content: sentences such as "the criminal shot the officer" and "the officer shot the criminal" are equally plausible responses to queries about criminals and officers, but much less equivalent from the perspective of law enforcement and the media. Solving the second problem, intelligence extraction, depends on understanding what a document is 'about' in a semantic sense, as well as aspects of each document's meta-properties: who wrote it, what their intent was in doing so, how it might be understood by an audience, whether it was intentionally deceptive, what attitudes and emotional tone it conveys, and a long list of other possibilities. In other words, many useful properties require understanding what might be called the social, or even sociological, properties of documents.

Determining such properties is key to domains such as e-discovery (finding significant emails in a corporate archive), intelligence (finding meaningful threats in a set of forum posts), measuring the effectiveness of a marketing campaign (in online social media posts), or predicting an uprising (using Twitter feed data).

Although bag-of-words approaches have been moderately successful for such problems, they tend to hit a performance wall (80% prediction accuracy is typical) because the representation fails to capture sufficient subtleties [30]. There have been attempts to increase the quality of representations, for example by extracting parse trees (that is, context-free grammar representations) but this focuses entirely on (somewhat artificial) language structure, and not at all on mental processes [14]. Other approaches leverage syntactically expressed semantic information, for example by counting word bigrams, by using Wordnet [26], or using deep learning [29]. Recent developments in deep learning, particularly LSTMs and biLSTMs have increased prediction accuracy; but these predictors are black boxes, so they bring little understanding of why and how a property is present in a document.

One approach that shows considerable promise is *systemic functional linguistics* [9, 12, 19], a model of language generation with sociological origins and an explicit focus on the effect of the creator's mental state and social setting on a created document. In this model, the process of generating an utterance (a sentence, a paragraph, or an entire document) is conceived of as traversing a *systemic net*, a set of structured choices. The totality of these choices defines the created document. At some nodes, the choice is disjunctive: continue by choosing *this* option or by choosing *that* one. At others, the choice is conjunctive: choose a subset of these options and continue in parallel down several paths.

Figure 1 shows a simple example of a systemic net. The decision to communicate requires a parallel (independent) choice of the level of formality to be used and the communication channel to be used. The level of formality could be formal or

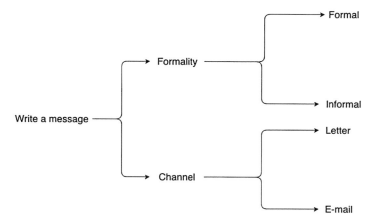

Fig. 1 A simple example of a systemic net

informal; and the channel could be via physical letter or email. These choices at the second level are disjunctive—it has to be one or the other. Further choices exist below these ones, so the systemic net notionally continues to the right until it results in concrete language.

Production using a context-free grammar also requires a structured set of choices, but the choices are top-down (so that the first choice is to instantiate, for example, a declarative sentence as a subject, an object, and a verb). In contrast, the order of the choices in a systemic net has no necessary relationship to the concreteness of the implications of those choices. For example, the choice to use formal or informal style is an early choice with broad consequences that limit the possibilities for subsequent choices. The choice to write a letter or an email is also an early choice but its immediate consequence is narrow and low level: typically whether the first word of the resulting document will be "Dear" (for a letter) or not (for an email).

Another example of a well-used systemic net, called the Appraisal Net [1], is shown in Fig. 2. It describes the way in which choices of adjectives are made when evaluating some object. The choice process is not arbitrary; rather an individual chooses simultaneously from up to three parallel paths: appreciation, affect, and judgement. Within two of these choices, there are then subsequent parallel choices that lead to particular adjectives—one example adjective is shown at each leaf. These choices are associated with different aspects of the situation: composition-complexity captures aspects of the object being appraised, while reaction-quality captures aspects of the person doing the appraising.

The power of systemic nets comes because these choices are made, not simply with the goal of constructing a syntactically valid sentence, but because of the limitations and exigencies of *social purpose* (certain things cannot be said in certain circumstances although syntactically valid); *mental state* (because language generation is a largely subconscious process), and the properties of the language in use. In other words, the choice of adjective in an appraisal certainly says something

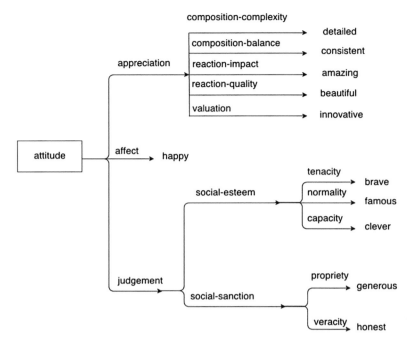

Fig. 2 The Appraisal systemic net, appropriate for representing judgements or reviews

about the object being assessed, but also reveals something about the person doing the assessing; and the structure of the choices would be different in English from, say, French or Japanese.

A systemic net is explanatory at three different levels. First, the existence of a net organizes constructions into categories and so explains some aspects of how the pieces in a text fit together.

Second, the choices made by individuals traversing a net are not typically unique; rather, they cluster into common choice patterns that reflect particular kinds of textual targets. This is because there are social rules that govern acceptable end-products. Each individual can write with an individual style, but we can also say that some set of documents by different authors are written in a down-to-earth style, and another set in a flowery style. This idea of a consistent set of choices in a net, leading to detectable consistencies in the resulting documents is called a *register*. Thus the set of registers associated with a net are also explanatory.

Third, for any particular document we can list the choices made in its construction, and this becomes a record that describes that document at a higher level of abstraction than as a bag of words. This level of explanation is most directly useful for analytics—such choices can be used as attributes for clustering or for prediction.

The advantages of a systemic functional approach to textual analytics are:

– The choices within the net are a smaller, more abstract, and more structured set than the choice of individual words, and therefore provide a stronger foundation for knowledge discovery—a kind of structured attribute selection; and
– These choices reflect, and make accessible, the mental state of the author or speaker and his/her perception of the social situation for which the text was constructed. This enables a kind of reverse engineering of how the text came to be, that is analytics about authors and settings.

The reason why systemic net approaches have not been more widely used in text analytics is because they have, so far, been constructed by computational linguists, often requiring several person-years to build, even when of modest size. Some substantial systemic nets have been built, but usually within the context of projects where they have been kept confidential; those that are public, like the Appraisal Net above, are usually small.

The contributions of this chapter are:

– We show that it is possible to infer systemic nets from corpora using Non-Negative Matrix Factorization (NNMF), and that these nets are plausible. Thus we are able to construct systemic nets for any corpus, and for any set of relevant words. This creates a new path to representing corpora at a deeper level, but without the need (and cost) for substantial human input.
– We show that the resulting systemic nets organize corpora more strongly than the corresponding bags of words, and that this organization improves both clustering and prediction tasks, using authorship prediction as a demonstration task.
– We apply systemic functional nets to a real-world intelligence problem, learning systemic nets from a set of Islamist magazines, and applying the resulting structure to two large Islamist forums. We show that the top-level distinctions derived from the magazines can also be clearly seen in the forum posts, suggesting a widespread mindset shared by the audience for these ideas.

2   Related Work

There have been several applications of predefined systemic nets to textual prediction problems. For example, Whitelaw et al. [32] show improvement in sentiment analysis using the Appraisal Net mentioned above. Argamon et al. show how to predict personality type from authored text, again using systemic functional ideas [31]. Herke-Couchman and Patrick derive interpersonal distance from systemic network attributes [13].

The most successful application of systemic functional techniques is the Scamseek project. The goal of this project was to predict, with high reliability, web pages that represented financial scams and those that represented legitimate financial products. This is a challenging problem—the differences between the two classes are small and subtle, and even humans perform poorly at the margins. The fraction of documents representing scams was less than 2% of the whole. This project's predictive model was successfully deployed on behalf of the Australian Securities and Investments Commission [21]. However, the effort to construct the registers corresponding to normal and (many varieties of) scam documents was substantial.

Kappagoda [15] shows that word-function tags can be added to words using conditional random fields, in the same kind of general way that parsers add part-of-speech tags to words. These word-function tags provide hints of the systemic-functional role that words carry. This is limited because there is no hierarchy. Nevertheless, he is able to show that the process of labelling can be partially automated and that the resulting tags aid in understanding documents.

Especially since the World Trade Center attacks of 2001, there has been a great deal of academic work on open-source intelligence [4, 17, 24]. This includes leveraging text [16, 25, 33] and graph data, including social networks [5, 6, 10, 22]. A large number of commercial platforms have also been developed and are in widespread use, for example i2 Analysts' Notebook and Palantir.

3   Inductive Discovery of Systemic Nets

The set of choices in a systemic net lead eventually, at the leaves, to choices of particular (sets of) words. One way to conceptualize a systemic net, therefore, is as a hierarchical clustering of words, with each choice representing selection of a subset.[1] We use this intuition as a way to inductively construct a systemic net: words that are used together in the same document (or smaller unit such as a sentence or paragraph) are there because of a particular sequence of choices. An inductive, hierarchical clustering can approximate a hierarchical set of choices.

Our overall strategy, then, is to build document-word matrices (where the document may be as small as a single sentence), and then cluster the columns (that is, the words) of such matrices using the similarity of the documents in which they appear. The question then is: which clustering algorithm(s) to use.

In this domain, similarity between a pair of documents depends much more strongly on the *presence* of words than on their *absence*. Conventional clustering algorithms, for example agglomerative hierarchical clustering and other algorithms that use distance as a surrogate for similarity, are therefore not appropriate, since mutual absence of a word in two different documents is uninformative, but still increases their apparent similarity.

Singular value decomposition is reasonably effective (J.L. Creasor, unpublished work) but there are major issues raised by the need to normalize the document-word matrix so that the cloud of points it represents is centered around the

---

[1]Complete systemic nets also include a downstream phase that defines the process for assembling the parts of a constructed document into its actual linear sequence. We ignore this aspect. In declarative writing, assembly is usually straightforward, although this is not the case in, for example, poetry.

origin. Typical normalizations, such as z-scoring, conflate median frequencies with zero frequencies and so introduce artifacts that are difficult to compensate for in subsequent analysis.

We therefore choose to use Non-Negative Matrix Factorization, since a document-word matrix naturally has non-negative entries. An NNMF decomposes a document-word matrix, A, as the product of two other matrices:

$$A = WH$$

If A is $n \times m$, then W is $n \times r$ for some chosen r usually much smaller than either m or n, and H is $r \times m$. All of the entries of W and H are non-negative, and there is a natural interpretation of the rows of H as 'parts' that are 'mixed' together by each row of W to give the observed rows of A [18].

Algorithms for computing an NNMF are iterative in nature, and the results may vary from execution to execution because of the random initialization of the values of W and H. In general, the results reported here are obtained by computing the NNMF 10 times and taking the majority configuration. We use a conjugate gradient version of NNMF, using Matlab code written by Pauca and Plemmons.

There are two alternative ways to use an NNMF, either directly from the given data matrix, or starting from its transpose. If we compute the NNMF of the transpose of A, we obtain:

$$A' = \bar{W}\bar{H}$$

and, in general, it is not the case that $\bar{H} = W'$ and $\bar{W} = H'$. Experiments showed that results were consistently better if we applied the NNMF to A', that is to the word-document matrix. The textual unit we use is the paragraph. A single sentence might, in some contexts, be too small; a whole document is too large since it reflects thousands of choices.

We extracted paragraph-word matrices in two ways. A parts-of-speech-aware tagger made it possible to extract the frequencies of, for example, all pronouns or all determiners [7]. For larger word classes, such as adjectives, it was also possible to provide the tagger with a given list and have it extract only frequencies of the provided words. Frequency entries in each matrix were normalized by the total number of words occurring in each paragraph, turning word counts into word rates. This compensates for the different lengths of different paragraphs.

Superior results were obtained by choosing only $r = 2$ components. In the first step, the $\bar{W}$ matrix has dimensionality *number of words* $\times$ 2, with non-negative entries. Each word was allocated to the cluster with the largest entry in the corresponding row of $\bar{W}$, and the process repeated with the two submatrices obtained by splitting the rows of A' based on this cluster allocation. This process continued until the resulting clusters could not be cleanly separated further. These clusters therefore form a binary tree where each internal node contains the union of the words of its two children.

Each NNMF was repeated 10 times to account for the heuristic property of the algorithm. We were able to leverage this to estimate the confidence of each clustering. For example, there were occasionally particular words whose membership oscillated between two otherwise stable clusters, and this provided a signal that they didn't fit well with either. We were also able to use this to detect when to stop the recursive clustering: either clusters shrank until they contained only a single word (usually a high-frequency one), or their subclusters began to show no consistency between runs, which we interpreted to mean that the cluster was being over-decomposed.

The result of applying this recursive NNMF algorithm to a word-paragraph matrix is a hierarchical binary tree whose internal nodes are interpreted as choice points, and whose leaves represent the 'outputs' that result from making the choices that result in reaching that leaf. A leaf consists of a set of words that are considered to be, in a sense, equivalent or interchangeable from the point of view of the total set of words being considered. However, this view of leaves contains a subtle point. Suppose that a leaf contains the words 'red' and 'green'. These are clearly not equivalent in an obvious sense, and in any given paragraph it is likely that an author will select only one of them. In what sense, then, are they equivalent? The answer is that, from the author's point of view, the choice between them is a trivial one: either could serve in the context of the document (fragment) being created. Thus a leaf in the systemic net contains a set of words from which sometimes a single word is chosen and sometimes a number of words are chosen—but in both cases the choice is unconstrained by the setting (or at least undetectably unconstrained in the available example data).

We have remarked that choices at internal nodes in a systemic net can be disjunctive or conjunctive. However, in our construction method each word in a particular document is allocated to exactly one cluster or the other. We estimate the extent to which a choice point is conjunctive or disjunctive by counting how often the choice goes either way across the entire set of documents, that is we treat conjunction/disjunction as a global, rather than a local, property. (It would be possible to allocate a word to both clusters if the entries in the corresponding row of $\bar{W}$ had similar magnitude, and therefore detect conjunctive choices directly. However, deciding what constitutes a similar magnitude is problematic because of the variation between runs deriving from the heuristic nature of the algorithm.)

4   Inferred Systemic Nets

The data used for proof of concept of this approach is a set of 17 novels downloaded from gutenberg.org and lightly edited to remove site-specific content. These novels covered a period of about a century from the 1830s to the 1920s and represent well-written, substantial documents. For processing they were divided into paragraphs; because of the prevalence of dialogue in novels, many of these paragraphs are actually single sentences of reported speech. The total number of paragraphs is

Table 1 List of words used to create the systemic networks

Group type	Words
Personal pronouns	I, me, my, mine, myself, we, us, our, ours, ourselves, you, your, yours, yourself, yourselves, they, their, theirs, them, themselves, he, him, his, himself, she, her, hers, herself, it, its, itself, one, one's
Adverbs	Afterwards, already, always, immediately, last, now, soon, then, yesterday, above, below, here, outside, there, under, again, almost, ever, frequently, generally, hardly, nearly, never, occasionally, often, rarely
Auxiliary verbs	Was, wasn't, had, were, hadn't, did, didn't, been, weren't, are, is, does, am, has, don't, haven't, doesn't, aren't, do, isn't, have, be, hasn't
Positive auxiliary verbs	Was, had, were, did, been, is, does, are, am, has, do, have, be
Adjectives	Good, old, little, own, great, young, long, such, dear, poor, new, whole, sure, black, small, full, certain, white, right, possible, large, fresh, sorry, easy, quite, blue, sweet, late, pale, pretty
Verbs	Said, know, see, think, say, go, came, make, come, went, seemed, made, take, looked, thought, saw, tell, took, let, going, get, felt, seen, give, knew, look, done, turned, like, asked

48,511. The longest novel contained 13,617 paragraphs (*Les Miserables*) and the shortest 736 (*The 39 Steps*).

We selected six different categories of words for experiments as shown in Table 1.

Figure 3 shows the systemic net of pronouns. In all of these figures, the thickness of each line indicates how often the corresponding path was taken as the result of a choice. Lines in blue represent the 'upper' choice, red the 'lower' choice, and black the situation where both choices occurred with approximately equal frequency.

The top-level choice (1) in this net is between pronouns where the point of view is internal to the story, and where the point of view is of an external narrator. This seems plausible, especially in the context of novels. Choice point 2 is largely between first-person and second-person pronouns, with apparently anomalous placement of 'me' and 'we'. Choice point 4 is between masculine pronouns and others, again entirely plausible given the preponderance of masculine protagonists in novels of this period. The remaining choices in this branch separate feminine, impersonal, and third-person plural pronouns. All of these choices are strongly

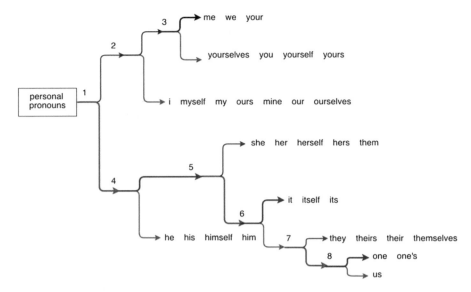

Fig. 3 Systemic net inferred for pronouns

disjunctive, weakening down the tree with choice point 7 the least disjunctive. It might be expected that, after the choice at point 1, choices might become more conjunctive as two or more people are mentioned. However, reported speech by one person is the most common paragraph structure in these novels, and many of these do not contain another pronoun reference ("He said 'What's for dinner?").

Figure 4 shows the systemic net for auxiliary verbs. These might have separated based on their root verb (to be, to have, to do)[2] but in fact they separate based on tense. Choice point 1 is between past tense forms and present tense forms. Choices between verb forms are visible at the subsequent levels. Of course, auxiliary verbs are difficult to categorize because they occur both as auxiliaries, and as stand-alone verbs.

The set of auxiliary verbs is also difficult because many of them encapsulate a negative ('hadn't'), and negatives represent an orthogonal category of choices. Figure 5 shows that systemic net when only the positive auxiliary verbs are considered. Again, tense is the dominant choice.

Figure 6 shows the systemic net for adverbs from a limited set of three different kinds: time, place, and frequency. This systemic net seems unclear, but note that at least some branches agree with intuition, for example the lower branch from choice four.

There are a very large number of adjectives used in the corpus, most of them only rarely. However, it is interesting to consider how adjectives might be empirically

---

[2] And a computational linguist might have chosen this separation as the most 'natural'.

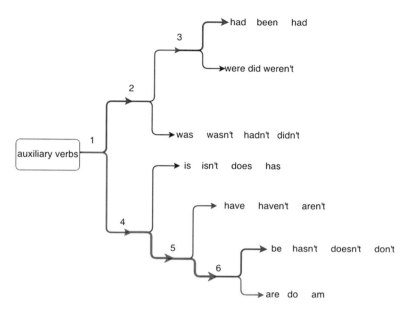

Fig. 4 Systemic net inferred for auxiliary verbs

Fig. 5 Systemic net inferred for positive auxiliary verbs

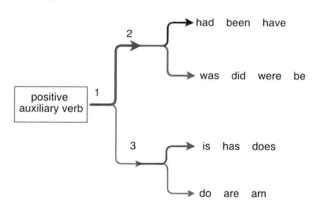

distinguished in fiction. (Note that this would not be the same net as the Appraisal Net described earlier, which might be inferrable from, say, a corpus of product reviews.) Figure 7 shows the systemic net for a limited set of adjectives of three kinds: appearance, color, and time. This net shows the typical structure for an extremely common word, in this case 'good' which appears as one outcome of the first choice. The sets of adjectives at each leaf are not those that would be conventionally grouped, but there are a number of interesting associations: 'great' and 'large' occur together, but co-occur with 'black' which is a plausible psychological association.

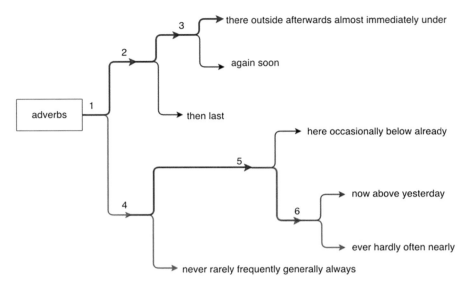

Fig. 6 Systemic net inferred for adverbs

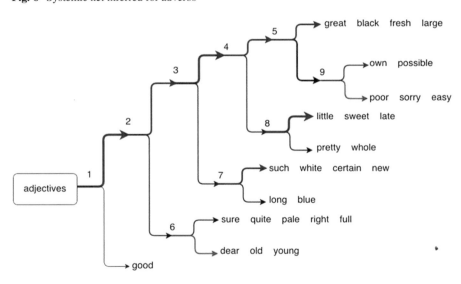

Fig. 7 Systemic net inferred for adjectives

These systemic nets look, from a human perspective, somewhere between plausible and peculiar. We now turn to more rigorous validation. Our goal is not so much that these nets should be explanatory from an intuitive perspective, but that they should be useful for analytic tasks (Fig. 8).

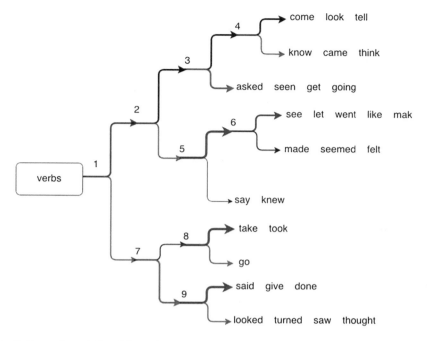

Fig. 8 Systemic net inferred for verbs

5 Validation

To validate our technique for inferring systemic nets, we use the following methods:

– Face validation. The systemic nets should involve choices that appear sensible and realistic. Note that this does not mean that they should match the hierarchy created to explain English grammar—such a grammar is an artificial construct intended to suggest consistent rules, and owing much to the grammar of Latin, rather than an accurate description of how English actually works.
– Comparison of document clustering based on word choices and based on systemic net choices. If choices reflect deeper structure, then documents should cluster more strongly based on choice structure than on word structure.
– Comparison of the performance of an example prediction task, authorship prediction, using word choices and systemic net choices. If choices reflect deeper structure, it should be easier to make predictions about documents based on choice structure than on word structure.
– Comparison with randomly created choice nets. Hierarchical clusterings with the same macroscopic structure as induced systemic nets should perform worse than the induced systemic nets.

5.1 Face Validation

The systemic nets shown in the previous section are not necessarily what a linguist might have expected, but it is clear that they capture regularities in the way words are used (especially in the domain of novels that was used, with their emphasis on individuals and their high rates of reported speech).

5.2 Clustering Using Word Choices Versus Net Choices

The difference between the systemic net approach and the bag-of-words approach is that they assume a different set of choices that led to the words that appear in each paragraph. The bag-of-words model implicitly assumes that each word was chosen independently; the systemic net model assumes that each word was chosen based on hierarchical choices driven by purpose, social setting, mental state, and language possibilities. Clustering paragraphs based on these two approaches should lead to different clusters, but those derived from systemic net choices should be more clear-cut. In particular, choices are not independent both because of hierarchy and because of the extrinsic constraints of the setting (novels, in this case)—so we expect to see clusters corresponding to registers.

We used two novels for testing purposes: *Robinson Crusoe* and *Wuthering Heights*, processed in the same way as our training data. Since these novels were not used to infer the systemic nets, results obtained using them show that the nets are capturing some underlying reality of this document class.

We compute the singular value decomposition of the paragraph-word matrix and the paragraph-choices matrix, both normalized by paragraph length. Plots show the resulting clustering of the paragraphs, with one test novel's paragraphs in red and the other in blue. In all of Figs. 9, 10, 11, and 12 the clustering derived from word frequencies is a single central cluster. In some of them, there appears to be

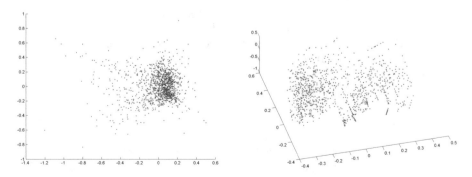

Fig. 9 SVD using pronouns, bag-of-words (left), choices (right)

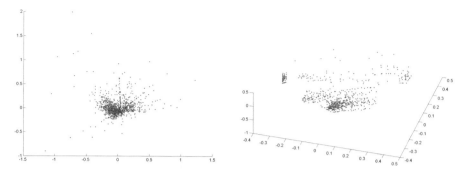

Fig. 10 SVD using auxiliary verbs, bag-of-words (left), choices (right)

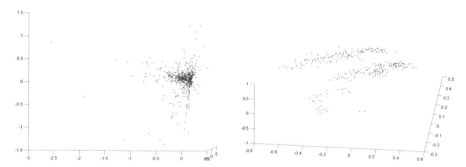

Fig. 11 SVD using adjectives, bag-of-words (left), choices (right)

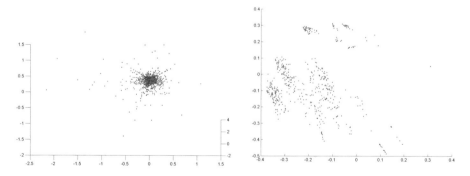

Fig. 12 SVD using verbs, bag-of-words (left), choices (right)

a separation between the two test documents, but these are illusions caused by overlays of points. In contrast, the clustering using choices shows strong clusters. These correspond to paragraphs that resulted from similar patterns of choices, that is to registers.

5.3 Authorship Prediction Using Word Choices Versus Net Choices

We argued that systemic nets are useful for applications where properties other than simple content are significant. To justify this claim we predict authorship *at the level of each individual paragraph* for our two test novels. This is a difficult task because paragraphs are so short; even humans would find it difficult to predict authorship at this level, especially without access to the semantics of the words used. Our goal is to show that the choice structure of the nets improves performance over simple use of bags of words. There are, of course, other ways to predict authorship, for example word n-grams or deep learning using LSTMs, but these are not directly comparable to systemic net approaches.

Again we use paragraph-word and paragraph-choice matrices as our data, and 5-fold cross-validated support vector machines with a radial basis kernel as the predictors. Results are shown for each of the word sets in Tables 2, 3, 4, 5, 6 and 7.

Across all of these word classes, authorship prediction based on word use hovers close to chance; in contrast, authorship prediction using systemic net choices range from accuracies of around 65%–75%, that is performance lifts of between 15 and 20 percentage points over prediction from word choices. And of these models is using only small numbers of words as signals of authorship. Clearly, the structural information coded in the systemic nets makes discrimination easier.

5.4 Inferred Nets Versus Randomly Generated Nets

Tables 8 and 9 compare the authorship prediction performance of the inferred systemic net and random networks constructed to have the same shape by dividing the words hierarchically into nested subsets of the same sizes as in the systemic net, but at random.

Table 2 Confusion matrices for personal pronouns; accuracy using words: 69.7%, accuracy using choices: 75.3%

Actual	Predicted: words and choices			
	RobCrusoe	WutHeights	RobCrusoe	WutHeights
RobCrusoe	694 (48%)	33 (2%)	584 (40%)	143 (10%)
WutHeights	407 (28%)	320 (22%)	216 (15%)	511 (35%)

Table 3 Confusion matrices for adverbs; accuracy using words: 51.3%, accuracy using choices: 63.4%

Actual	Predicted: words and choices			
	RobCrusoe	WutHeights	RobCrusoe	WutHeights
RobCrusoe	171 (12%)	556 (38%)	387 (27%)	340 (23%)
WutHeights	152 (10%)	575 (40%)	192 (13%)	535 (37%)

Table 4 Confusion matrices for auxiliary verbs; accuracy using words: 50.6%, accuracy using choices: 72.0%

Actual	Predicted: words and choices			
	RobCrusoe	WutHeights	RobCrusoe	WutHeights
RobCrusoe	435 (30%)	292 (20%)	553 (38%)	174 (12%)
WutHeights	426 (29%)	301 (21%)	233 (16%)	494 (34%)

Table 5 Confusion matrices for positive auxiliary verbs; accuracy using words: 51.4%, accuracy using choices: 67.6%

Actual	Predicted: words and choices			
	RobCrusoe	WutHeights	RobCrusoe	WutHeights
RobCrusoe	453 (30%)	292 (20%)	623 (43%)	104 (7%)
WutHeights	415 (29%)	312 (21%)	367 (25%)	360 (25%)

Table 6 Confusion matrices for adjectives; accuracy using words: 50.1%, accuracy using choices: 70.8%

Actual	Predicted: words and choices			
	RobCrusoe	WutHeights	RobCrusoe	WutHeights
RobCrusoe	295 (20%)	432 (30%)	490 (34%)	237 (16%)
WutHeights	294 (20%)	433 (30%)	187 (13%)	540 (37%)

Table 7 Confusion matrices for verbs; accuracy using words: 50.1%, accuracy using choices: 67.5%

Actual	Predicted: words and choices			
	RobCrusoe	WutHeights	RobCrusoe	WutHeights
RobCrusoe	297 (20%)	430 (30%)	476 (33%)	251 (17%)
WutHeights	296 (20%)	431 (30%)	221 (15%)	506 (35%)

Table 8 Personal pronouns: systemic network versus random nets

Number of paragraphs	NNMF systemic network	Random nets		
	Accuracy	min	mean	max
1	75.3%	69.3%	75.4%	82%
3	84.1%	68.1%	72.1%	76.2%
6	88.9%	66.3%	70%	71.5%

Table 9 Adjectives systemic network versus random nets

Number of paragraphs	NNMF systemic network	Random nets		
	Accuracy	min	mean	max
1	70.8%	70.2%	75.2%	79.4%
3	72.3%	66.9%	71.5%	73%
6	74.8%	64.5%	69.4%	72.3%

The performance of the random network is approximately the same as the inferred network at the level of single paragraph prediction. This is clearly a small sample size effect: choices that differentiate authors well are also available in the random network by chance. However, as the number of paragraphs available to make the prediction increases, the predictive performance of the systemic net continues to improve while that of the random network remains flat.

5.5   Combining Systemic Nets

We have built our systemic nets starting from defined word sets. In principle, a systemic net for all words could be inferred from a corpus. However, such a net would represent, in a sense, the entire language generation mechanism for English, so it is unlikely that it could be reliably built, and would require an enormous corpus.

However, it is plausible that the systemic nets we have built could be composed into larger ones, joining them together with an implied conjunctive choice at the top level. We now investigate this possibility.

One way to tell if such a composition is meaningful is to attempt the authorship prediction task using combined systemic nets. The results are shown in Table 10. The combined nets show a lift of a few percentage points over the best single net.

These results hint, at least, that complex systemic nets can be built by inferring nets from smaller sets of words, which can be done independently and perhaps robustly; and then composing these nets together to form larger ones. Some care is clearly needed: if the choice created by composing two nets interacts with the choices inside one or both of them, then the conjunctive composition may be misleading. This property is known as selectional restriction, and is quite well understood, so that it should be obvious when extra care is needed. For example, composing a net for nouns and one for adjectives using a conjunctive choice is unlikely to perform well because the choice of a noun limits the choice of adjectives that 'match' it.

Table 10 Prediction accuracy using combined word sets, best single systemic network, and combinations of systemic networks

	Words	Best single	Combined
Pronouns + adverbs	69%	75.3%	77.4%
Pronouns + adverbs + verbs	73.1%	75.3%	80.2%
Pronouns + adverbs + verbs + adjectives	80.37%	75.3%	80.44%

Table 11 Salafist-Jihadist words

Arabic	أَمريكا	ر ب	ظالمين	رسول	عدو	يهود
English	America	God	Oppressors	Prophet	Enemy	Jews

6 Applying Systemic Nets for Intelligence Analysis

We now turn to applying the systemic net construction technique to text datasets that have intelligence value, documents created with islamist purposes of varying intensity.

A model of jihadist intensity developed by Koppel et al. [8] was designed to distinguish different strands of Islamic thoughts. We use the model (or set of words) describing a Salafist-jihadist orientation. The words were originally in Arabic, and were translated into English by the second author, a native Arabic speaker. The resulting list consists of 144 words. Table 11 shows some of words.

We show that a structure derived from three English-language islamist propaganda magazines, *Inspire*, *Dabiq* and *Azan*, generalizes elegantly to two islamist forums, showing that there is a widely held mindset (or set of distinctions) shared in this worldwide community.

7 Measuring Islamist Language

One way to measure the jihadi intensity of a given document is simply to sum the frequency with which relevant words occur. This approach has been widely used by, for example, Pennebaker to measure a number of properties [11] and the LIWC package has made this technology available to many researchers. This approach has been used to measure deception [20], informative language [23], imaginative language [23], and propaganda [3, 27, 28].

We examine three different jihadist magazines, all with the same professed goals, but originating from different countries, and from groups with different ideologies. Inspire is produced by Al Qaeda in the Arabian Peninsula. The first nine issues were edited by the American jihadist, Anwar al-Awlaki. Since his death, others, so far unidentified, have taken over. Dabiq is produced by ISIS (Daish) in Syria. Azan is produced by the Taliban in Pakistan. All three types of magazines are high in production values, use many visual images, and aim to imitate the look and feel of mainstream Western magazines. All of these magazines appear as pdfs; more recent issues have become so complex that it is impossible to extract the textual content using OCR but the text of 12 issues of Inspire, 5 issues of Azan, and 5 issues of Dabiq have been extracted. Skillicorn used the magazine data to do an empirical assessment of the intensity of propaganda across the three magazines [3].

A document-word matrix for the magazines was constructed by counting the frequencies of words from the jihadi language model, normalized as discussed

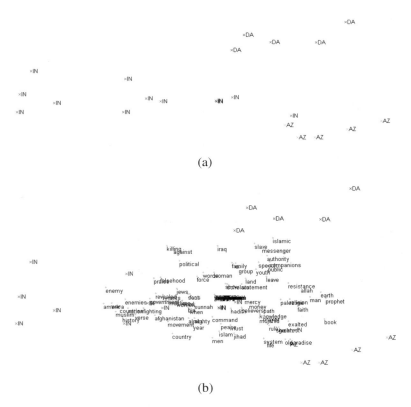

(a)

(b)

Fig. 13 Similarity among magazines using SVD (IN: Inspire, AZ: Azan, DA: Dabiq. (**a**) A plot of the magazines based on the jihadi language model. (**b**) A plot of the magazines based on the jihadi language model overlaid with the words, i.e. the U and V matrices plotted together

above. The SVD plots of the document-word matrix based on the jihadi words are shown in Fig. 13. Both Azan and Dabiq cluster strongly. Inspire does not cluster as well, suggesting that it does not have a consistent style or content focus. Figure 13b shows the magazines overlaid with the words they use—words and magazines can be considered to be pulled towards one another whenever a particular word is heavily used in a particular magazine issue. Some of the words most associated with Dabiq, therefore, are: 'Iraq', 'Islamic', 'slave', and 'authority'. This seems reasonable since ISIS is active in Syria and Iraq. ISIS also allows the practice of slavery. Some of the words most associated with Azan are: 'Paradise', 'life', 'old', and 'system'. These words suggest that Azan's message is more focused towards the afterlife.

8 A Systemic Net Based on Jihadi Language Model

We use the magazines to construct a systemic net derived from jihadi language. Figure 14 shows the resulting systemic net labelled with the choice points for later reference. The tree has six choice points and six leaves, because the words at node 7 do not split further.

The first choice in the tree reflects a clear distinction between political and religious words. A sample of the words associated with this choice is given in Table 12.

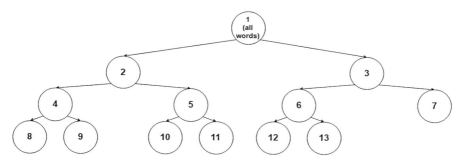

Fig. 14 A visualization of the systemic net

Table 12 First choice words

First choice	
Religious	Political
Prophet	Tyrants
Sheikh	Enemy
God	Fighting
Worlds	America
Paradise	Brotherhood
Exalted	Government
Earth	Palestinian
Monotheism	Oppressors
Goodness	Nation
Platform	Categories
Islamic	Act
Faith	Movement
Old	Nation
Owners	War
Allah	Doubt
Allowing	Country
Prayer	Ruler
Companions	Afghanistan

Table 13 Basic meaning behind each leaf node and the Google results obtained from using the words in each node

Leaf	Meaning	Google results
8	Used by inspire the most	Verses of Quran on Jihad–Islam
9	Used by Dabiq the most	Allah's Quran—authenticity of the Quran
		Muhammad, Terrorist or Prophet?—Bible Probe
		Islam and antisemitism—Wikipedia, the free encyclopedia
		what every non-Muslim needs to know about Islam!—Bible.ca
10	Jihad focused	Islamic State and the Others · Raqqa is Being Slaughtered
		How Islam will dominate the world — - Duaat - WordPress.com
		Chapter 1: Muhammad and the Quran
11	Pure religion	Prophet Muhammad, pbuh - Some selected verses
		Does Islam regard non - Muslims with mercy and compassion
		The Book of Faith - Sahih Muslim - Sunnah.com
12	Al-qaeda and Afghanistan focused	Islam in Afghanistan - Wikipedia, the free encyclopedia
		Afghan Arabs - Wikipedia, the free encyclopedia
		Al-Qaeda - Infoplease
13	Teachings about Islam	Contemporary Islamist ideology authorizing genocidal murder
		Full text of "Islamic Books by Ibn Taymiyyah Maqdisi"
		Do the authentic teachings of Islam result in terrorism?
		Welcome to IONA masjid and learning center!—IONA Masjid !

The word sets resulting from some of the choices make immediate intuitive sense, while others do not. As a way to understand what each choice is capturing, we take the word sets from each leaf, and treat them as terms for a Google query. The top ranked documents associated with each of them are shown in Table 13. Most sets can be assigned a plausible meaning based on these results, the exception being node 8.

We can now compare how the magazines cluster based on bag of words versus based on choice sets. An SVD plot of the variation between the magazines based on choices is shown in Fig. 15. Compared to Fig. 13, Azan and Dabiq cluster more tightly based on choices than on words. There is no significant difference between the two approaches in how Inspire issues cluster. The color and shape coding of the magazines is based on which choices are made most often at choice points in the systemic net. Inspire tends to prefers the political branch over the religious branch, but does not cluster well. This reflects the wide variation in focus that has been previously noted, and perhaps the changes in editorship and authorship. Both Azan and Dabiq have consistent choice patterns across all issues, suggesting clarity of purpose, and a consistent editorial framework. Choices 7 and 9 are strongly associated with Dabiq, while Inspire tends to favor the opposite choices (6 and 8).

This analysis allows us to associate particular sets of magazines with particular patterns of word choice, and therefore to take a first step towards judging group

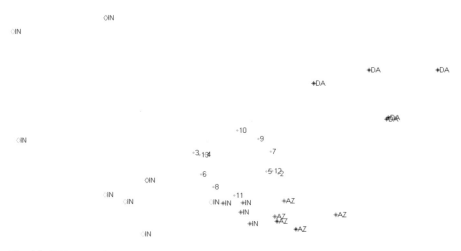

Fig. 15 SVD plot of the document-choice matrix of the magazines (diamonds: prefer political choice; asterisks: prefer religious choice; green and blue distinguish the outcome from choice point 2; black and red distinguish the outcome from choice point 3; numbers are the positions corresponding to the choices, i.e. the columns of the matrix)

intent. For example, it is possible to infer, in principle, whether a group's focus is internal or external, and if external what kind of target is likely to seem most attractive. This goes deeper than simply observing which words are frequent, because it associates words that are related in the sense that they occupy the same mental 'slot'.

9 Applying the Systemic Net to Islamist Forums

We apply the jihadi language systemic net that was inferred from the magazines to two new corpora, two islamist forums:

- Turn to Islam: which advertises itself as "correcting the common misconceptions about Islam".
- Islamic Awakening: which identifies itself as "dedicated to the blessed global Islamic awakening".

Turn to Islam (TTI) consists of 335,388 posts from 41,654 members collected between June 2006 and May 2013. Islamic Awakening (IA) consists of 201,287 posts from 3964 members collected between April 2004 and May 2012. Both data sets were collected by the University of Arizona Artificial Intelligence Lab [2]. The posts are primarily in English, but with a mixture of transliterated Arabic, some French, and a small number of words from other European languages.

Figure 16a shows the variation among a 10% uniformly random sample of posts from the TTI forum, based on the Jihadi language systemic net choices. The color and shape coding is the same as in Fig. 15. We can see a clear separation between posts making political versus religious word choices (diamonds vs asterisks). The separation also extends to the choice points in the second layer of the SFL net (blue/green and red/black). Figure 16b shows the cloud of points rotated so that the third dimension is visible, showing that the variation between red and black points is orthogonal to the variation between blue and green points. The striking point is that the choices inferred from word usage in the islamist magazines strongly and consistently cluster forum posts coming from a completely different context.

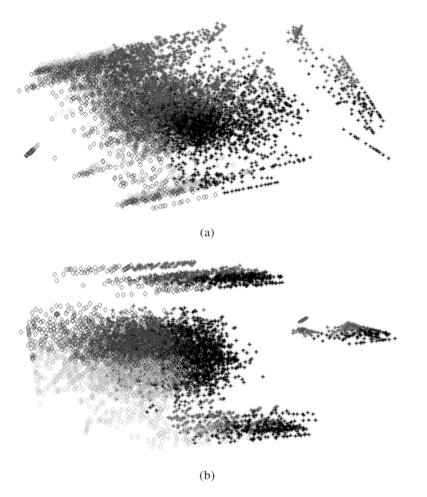

(a)

(b)

Fig. 16 SVD plot of the document-choice matrix of TTI posts (symbol and color coding as in Fig. 15). (**a**) First two dimensions. (**b**) Rotated view to show the third dimension

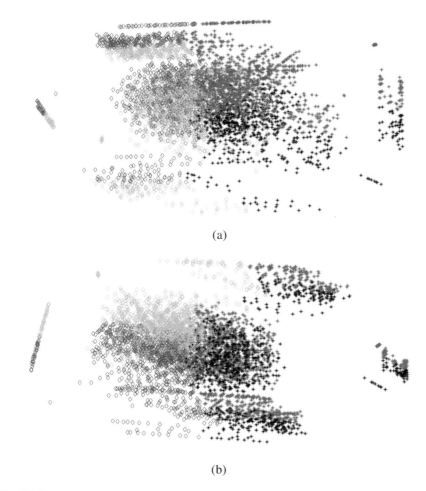

(a)

(b)

Fig. 17 SVD plot of the document-choice matrix of IA posts with the same symbol and color coding as Fig. 15. (**a**) First two dimensions. (**b**) Rotated view to show the third dimension

This suggests that there is a widely shared mindset in this community, interpreted broadly, that produces consistent language use across settings.

Figure 17 is the same analysis for the IA forums. Both TTI and IA cluster strongly and consistently based on the choice structure of the systemic net inferred from the magazines.

The word sets that result from choice point 5 are particularly interesting; they distinguish between two types of religious thinking, one that might be called purely religious and the other which is focused more on the jihad aspect of religion. The relevant words are shown in Table 14.

Figure 18 shows an SVD plot of TTI posts color-coded by Jihadi intensity which we obtain by adding two artificial documents that contain all of the words of the

Table 14 Choice point 5
words

Fifth choice point	
Pure religion	Jihad focused
Old	Behalf
Command	Hide
Exalted	Islamic
Folk	Monotheism
Mohammed	Authority
Allah	Owners
Peace	Jews
Believers	Resistance
Goodness	Companions
Faith	Woman
	Family
	Worlds
	Earth

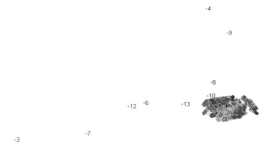

Fig. 18 SVD plot of TTI document-choice matrix. The posts are color coded based on jihadi intensity from blue (least intense) to red (most intense)

model at frequencies one standard deviation above, and one standard deviation below the mean, and using the line between them as a gradient of jihadi intensity. Blue points are the least jihadi and red are the most. Choice points from the systemic net are also included, and it is clear that the word sets that distinguish jihadi intensity most directly are those at nodes 10 and 11. Therefore, the choice made at choice point 5 could be also used, by itself, as a predictor of jihadi intensity.

10   Discussion

The strength of the religious vs political choice at the first choice point of the systemic net suggests that there is a fundamental differentiator among those who engage in islamist discussions or writings. It appears that islamist ideology can be plausibly separated into two threads, and this generalizes over different contexts and widely differing authors. It remains an open question whether the authors themselves are consciously aware of this; and whether an understanding of the dichotomy could be leveraged to increased the effectiveness of propaganda vehicles such as the magazines (and, for some posters, the forums). There is also a strong distinction in the religious domain between word choices that are, as it were, purely religious and those that are religious but with a jihadist subtext.

We have also demonstrated the effectiveness of systemic nets, and the choices they capture. The structure inferred from large, well-written islamist magazines generalizes very well to a completely different domain: short, informal posts in online forums.

Methodologically, we have shown that inferring systemic nets from data produces structures that reflect underlying language patterns, even though the word choice sets do not necessarily have a direct interpretation. The ability to infer systemic nets automatically, even if they are possibly not as accurate as those inferred by humans, opens up the SFL approach to many more application domains, of which intelligence analysis is just one.

References

1. S. Argamon, S. Dhawle, M. Koppel, J.W. Pennebaker, Lexical predictors of personality type, in *Proceedings of the Joint Annual Meeting of the Interface and Classification Society of North America* (2005)
2. Azure, Dark web forums. http://www.azsecure-data.org/dark-web-forums.html. AZSecure-data.org version (Accessed May 4th, 2016)
3. H. Borko, M. Bernick, Automatic document classification. J. ACM **10**(3), 151–162 (1963)
4. H. Chen, F.-Y. Wang, Artificial intelligence for homeland security. IEEE Intell. Syst. **20**(5), 12–16 (2005)
5. T. Coffman, S. Greenblatt, S. Marcus, Graph-based technologies for intelligence analysis. Commun. ACM **47**(3), 45–47 (2004)
6. D. Cook, L.B. Holder, Graph-based data mining. IEEE Intell. Syst. **15**(2), 32–41 (2000)
7. J.L. Creasor, D.B. Skillicorn, *QTagger: Extracting Word Usage from Large Corpora* (Queen's University, School of Computing, Kingston, 2012). Technical Report 2012-587
8. G.S. Davidson, B. Hendrickson, D.K. Johnson, C.E. Meyers, B.N. Wylie, Knowledge mining with VxInsight : discovery through interaction. J. Intell. Inf. Syst. **11**(259-285) (1998)
9. E.C. Davies, A retrospective view of Systemic Functional Linguistics, with notes from a parallel perspective. Funct. Linguistics **1**(1), 4 (2014)
10. J. Galloway, S. Simoff, Network data mining: discovering patterns of interaction between attributes, in *Advances in Knowledge Discovery and Data Mining*. Springer Lecture Notes in Computer Science, vol. 3918 (2006), pp. 410–414
11. Google WebAPI (2004). www.google.com/apis
12. M.A.K. Halliday, J.J Webster, *Bloomsbury Companion to Systemic Functional Linguistics*. Continuum Companions (Bloomsbury Academic, London, 2009)
13. M. Herke-Couchman, J. Patrick, Identifying interpersonal distance using systemic features, in *Proceedings of AAAI Workshop on Exploring Attitude and Affect in Text: Theories and Applications* (Springer, Netherlands, 2004), pp. 199–214

14. H. Kanayama, T. Nasukawa, H. Watanabe, Deeper sentiment analysis using machine translation technology, in *Proceedings of the 20th International Conference on Computational Linguistics* (2004)

15. A. Kappagoda, The use of systemic-functional linguistics in automated text mining, in *DSTO Defence Science and Technology Organisation DSTO-RR-0339* (2009). Technical report

16. C.E. Lamb, D.B. Skillicorn, Detecting deception in interrogation settings, in *IEEE International Conference on Intelligence and Security Informatics* (2013), pp. 160–162

17. M. Lazaroff, D. Snowden, Anticipatory models for counter-terrorism, in *Emergent Information Technologies and Enabling Policies for Counter-terrorism*, ed. by R.L. Popp, J. Yen, chapter 3, IEEE Press Series on Computational Intelligence (2006), pp. 51–73

18. D.D. Lee, H.S. Seung, Learning the parts of objects by non-negative matrix factorization. Nature **401**, 788–791 (1999)

19. C. Matthiessen, M.A.K. Halliday, *Systemic Functional Grammar: A First Step into the Theory* (Macquarie University, Macquarie Park, 1997)

20. M.L. Newman, J.W. Pennebaker, D.S. Berry, J.M. Richards, Lying words: predicting deception from linguistic styles. Personal. Soc. Psychol. Bull. **29**(5), 665–675 (2003)

21. J. Patrick, The scamseek project—text mining for financial scams on the internet, in *Data Mining: Proceedings of the 4th Australasian Workshop on Data Mining*. Springer LNCS 3755 (2006), pp. 295–302

22. J. Qin, J. Xu, D. Hu, M. Sageman, H. Chen, Analyzing terrorist networks: a case study of the global Salafi Jihad network, in *Intelligence and Security Informatics, IEEE International Conference on Intelligence and Security Informatics, ISI 2005, Atlanta, GA, USA, May 19-20*. Lecture Notes in Computer Science LNCS 3495 (Springer, Berlin, 2005), pp. 287–304

23. P. Rayson, A. Wilson, G. Leech, Grammatical word class variation within the British National Corpus sampler. Lang. Comput. **36**(1), 295–306 (2001)

24. M. Sageman, *Understanding Terror Networks* (University of Pennsylvania Press, Philadelphia, 2004)

25. A.P. Sanfilippo, A.J. Cowell, S.C. Tratz, A.M. Boek, A.K. Cowell, C. Posse, L.C. Pouchard, Content analysis for proactive intelligence: marshalling frame evidence, in *Proceedings of the Twenty-Second AAAI Conference on Artificial Intelligence* (2006), pp. 919–924

26. S. Scott, S. Matwin, Text classification using WordNet hypernyms, in *Natural Language Processing Systems: Proceedings of the Conference. Association for Computational Linguistics Somerset, New Jersey* (1998), pp. 38–44

27. D.B. Skillicorn, Lessons from a jihadi corpus, in *2012 IEEE/ACM International Conference on Advances in Social Networks Analysis and Mining (ASONAM)* (IEEE, Piscataway, 2012), pp. 874–878

28. D.B. Skillicorn, E. Reid, Language use in the jihadist magazines Inspire and Azan. Secur. Inform. **3**(1), 9 (2014)

29. R. Socher, A. Perelygin, J.Y. Wu, J. Chuang, C.D. Manning, A.Y. Ng, C. Potts, Recursive deep models for semantic compositionality over a sentiment treebank, in *Proceedings of the 2013 Conference on Empirical Methods in Natural Language Processing* (2013), pp. 1631–1642

30. Y.R. Tausczik, J.W. Pennebaker, The psychological meaning of words: LIWC and computerized text analysis methods. J. Lang. Soc. Psychol. **29**, 24–54 (2010)

31. C. Whitelaw, S. Argamon, Systemic functional features in stylistic text classification, in *Proceedings of AAAI Fall Symposim on Style and Meaning in Language, Art, Music, and Design, Washington, DC* (2004)

32. C. Whitelaw, N. Garg, S. Argamon, Using appraisal taxonomies for sentiment analysis, in *Second Midwest Computational Linguistic Colloquium (MCLC 2005)* (2005)

33. Y. Zhang, S. Zeng, L. Fan, Y. Dang, C.A. Larson, H. Chen, Dark web forums portal: Searching and analyzing jihadist forums, in *IEEE International Conference on Intelligence and Security Informatics* (2009), pp. 71–76

Twitter Bots and the Swedish Election

Johan Fernquist, Lisa Kaati, Ralph Schroeder, Nazar Akrami, and Katie Cohen

Abstract In this chapter, we present a study of how political Twitter bots were used before the Swedish general election in 2018. We have not restricted our study to bots that are a software program instead, we are interested in any type of bot-like automated behavior. This includes a human that manually copies or retweets content repeatedly in a robot-like way to influence the interaction between a user and content or with other users.

Our results show that bots were more likely to express support towards the immigration-critical party the Sweden Democrats, compared to genuine accounts.

Keywords Bots · Automated behavior · Twitter · Social media analysis · Election

1 Introduction

The Swedish general election was held on September 9th 2018 and was surrounded by great concern regarding the role of disinformation disseminated via digital media. The background to these concerns is broader debates about the role of digital media in politics that have intensified since the Brexit referendum and the election of Donald Trump. The Internet and social media are powerful tools for influencing political campaigns and discussions. Since 2016, the Internet has been used more than TV and newspapers as a source of political information in Sweden [17]. Ever

J. Fernquist · L. Kaati (✉) · K. Cohen
Swedish Defence Research Agency, Kista, Sweden
e-mail: johan.fernquist@foi.se; lisa.kaati@foi.se; katie.cohen@foi.se

R. Schroeder
Oxford University, Oxford, England
e-mail: nazar.akrami@psyk.uu.se

N. Akrami
Uppsala University, Uppsala, Sweden
e-mail: nazar.akrami@psyk.uu.se

© Springer Nature Switzerland AG 2020   141
M. A. Tayebi et al. (eds.), *Open Source Intelligence and Cyber Crime*, Lecture Notes
in Social Networks, https://doi.org/10.1007/978-3-030-41251-7_6

since the Brexit Referendum and Donald Trump's presidential election campaign in 2016, there has been a discussion about the role that digital media play in disinformation and influence operations in political campaigns. One precondition for effectively being able to counteract disinformation is a better understanding of how disinformation and attempts to influence politics work. What are the messages? How are they spread, and with what intentions?

In Sweden, the discussion about disinformation has focused on several areas: the use of alternative or partisan websites, the possibility of the spread of messages by foreigners and specifically Russia, and the use of bots to spread messages. However, bots are not only used for spreading disinformation. Bots have been used for a variety of purposes. While they were initially designed to automate otherwise unwieldy online processes which could not be done manually, they have come to be most commonly used for commercial purposes such as directing Internet users to advertisements and the like. Bots are also often used to further illegal activity such as collecting data from users for criminal gain.

There are many different definitions of 'bots'. In [13] bots have been defined as "executable software that automates the interaction between a user and content or other users". In the work by Gorwa and Guilbeault [11] a typology of bots is presented. The topology suggests six different types of bots:

- web robots (crawlers and scrapers)
- chatbots (human-computer dialog system which operates through natural language via text or speech)
- spambots (bots that post on online comment sections and spread advertisements or malware on social media platforms)
- social bots (various forms of automation that operate on social media platforms)
- sock puppets and "trolls" (fake identities used to interact with ordinary users on social networks)
- cyborgs and hybrid accounts (a combination of automation and human curation)

Web robots do not interact with users on a social platform and are therefore considered to be different from automated social media accounts. Social bots are bots that generally act in ways that are similar to how a real human may act in an online space. Social bots that are used for political purposes are called political bots. The term sock puppet refers to fake identities used to interact with ordinary users on social networks. Politically motivated sock puppets, especially when coordinated by governments or interrelated actors, are according to [11], called "trolls". According to the typology presented in [11], the bots we are studying in this work are both automated social bots and sock puppets. The aim with the definition of bots that we use is to capture an automated behavior. The automated behavior can be induced by humans working at a state-owned agency to spread propaganda (sock puppets), by fully automated software bots (spambots, social bots), and by humans that are behaving in an automated fashion (spambots). The effect of an automated behavior is the same independently of if it is a software or a human that is spreading the messages.

Political bots are used in various ways when they aim at influencing public opinion. For example, bots can be used to spread disinformation to mislead about the state-of-affairs. They can also be used to spread false news with the aim of creating uncertainty about established sources of information. Another aim of using bots is to lead users to think that specific content is more shared, more generally accepted or more mainstream than is the case. We will use *bot* and *automated account* interchangeably. As mentioned before, we do not define an automated account based on whether there is a human or a piece of software that produces the content but rather base this on the account's behavior.

Even tough bots can exist on many different platforms, we only focus on studying bots on Twitter. One of the reasons for studying Twitter is because it is a widely used public forum for political discussion in Sweden, especially among journalists [12]. In the rest of this chapter, we will present an analysis of how political bots were used on selected hashtags on Twitter before and after the Swedish general election in 2018. The data was collected between March 5th and September 30th.

The analysis of automated accounts presented in this chapter consist of three different parts:

- The number of automated accounts (bots)
- An analysis of the domains bot links to and what kind of messages they distribute
- How the bots communicate with other accounts on Twitter

1.1   Swedish Politics

In Sweden, a party must receive at least 4% of the votes in an election to be assigned a seat in the Swedish parliament. In 2018 eight parties received more than 4% of the votes. The largest party is the Social Democratic Party (S). S is a labor party at its core with policies based on freedom, equality, and solidarity. The party prioritizes the creation of more jobs and to provide a better education for all.

The Moderate Party (M) is the second largest party. M is a conservative party with liberal ideas. The individual's freedom to choose is central to its policies, and the party generally supports reduced taxes and economic liberalism.

The Sweden Democrats (SD) is a social conservative party based on nationalistic values. The party is associated with issues of migration, and the party's policies are based on protecting the *national identity* as a way of sustaining the Swedish welfare state.

The Centre Party (C) is a liberal and agriculture political party. The party believes that society should be built on people's responsibility for each other and nature and focus on the national economy, the environment, and integration.

The Left Party (V) defines itself as a socialist and feminist political party with an ecological basis. Focus areas are jobs, welfare services, and gender equality. The party was against Sweden joining the EU in 1995 and still advocates an exit. The Christian Democratic Party (KD) believes that stable families should form the basis of society. The four main issues that the Christian Democrats focuses on are: improving elderly care, giving families with children the freedom to select desired childcare, simplifying regulations for companies and lowering taxes as a means to promote growth and combat unemployment.

The Liberals (L) are a liberal and social—liberal political party that holds a middle position in the Swedish political landscape. The Green Party (MP) has a clear focus on environmental issues. The party focuses on stopping climate change and protecting the environment, fighting nuclear power and promoting European integration.

Apart from the eight parties that are in the Swedish parliament, there are also several smaller parties. Two newly formed parties that appear in our analysis are Alternative for Sweden (AFS) and Citizens' Coalition (MED). AFS was founded in 2017. The party's policies are based on immigration issues, democracy and politicians, and law and order. MED considers itself liberal-conservative and green conservative. Both parties are anti-immigration.

2   Method

In our analysis of the Swedish election, we use machine learning to detect accounts with automatic behavior. In the rest of this section, we describe how the classification model is built, how it performs compared to other bot detection models, and what data we have used to train the model. Our work is also put into relation to previous work in the domain.

2.1   *Classification of Bots*

We have trained a classification model to identify accounts exhibiting automatic behavior. The classification problem is to determine if an account is *genuine* or a bot. Here, a genuine account is an account that is operated by a "normal" human being. To build a model that is able to recognize automatic behavior, labeled training data is needed. The training data consists of accounts that are already known as bots or genuine accounts. When training our model, we use a number of different features. Our model is language independent but we have used it to classifying tweets in Swedish. The classification is described in more detail in [9].

Related Work

There have been several efforts dedicated to bot detection on Twitter. Random forest is the classification algorithm that has been proven to give the best performance for bot detection for the supervised problem when several different classifiers have been tested [14, 18, 19]. In [10] user meta-features and tweet features were used when training a classification algorithm. Their results indicated that bots have more URLs in their tweets and that they have a higher *follower-friend ratio*. The terminology used is from the Twitter API, where 'friend' indicates the number of users that the user is following, as opposed that the number of followers he or she has. In [10] it is also shown that genuine accounts get more likes on their tweets than bots.

In [2] bots and cyborgs are studied. The author states that follower-friend ratio might be a bad feature since the bots might be able to unfollow accounts which not are following them back automatically. Instead, they introduce text entropy as a feature to measure the similarity of the texts posted by an account with the hypothesis that bots have more uniform content in their tweets. Another feature that is considered is what kind of devices the different accounts are using when tweeting. Most of the genuine accounts are using the web or the mobile application while bots are using other applications such as the API. It was also noted that genuine accounts have a more complex timing behavior compared to bots and cyborgs.

In [19] a total of 1150 different features are used to train a model that recognizes bots. One set of features that is used is time features, including the statistics of times between consecutive tweets, retweets, and mentions. The results show that the two most informative feature types are user meta-data and content features. The content features include frequency and proportion of part-of-speech-tags (POS), number of words in a tweet and entropy of words in a tweet.

Training and Testing Data

We have used a number of different datasets to train our classification model. The first dataset was originally crawled during October and November 2015 and is described in [19]. The dataset contains labeled information about 647 bot accounts and 1367 genuine accounts. Each account has produced at least 200 tweets, of which at least 90 occurred during the crawling period. The accounts were manually annotated as bots or genuine. The annotation was based on characteristics such as profile appearance, produced content, and the interaction with other profiles.

The second dataset consists of 591 bots and 1680 genuine accounts [4]. The genuine accounts are Italian users that through a survey accepted to be a part of the study or accounts that were regularly active for a long period. The bot accounts were bought from a bot-service provider.

The third dataset was manually annotated by four undergraduate students [10]. The users in the dataset were divided into four subsets depending on the number of followers. The subsets were divided into users with more than 10 million followers, users with between 900 thousand and 1, 1 million followers, users with 90 thousand to 110 thousand followers and users with 900 to 1100 followers. We only use the two sets with users with 90 thousand to 110 thousand followers and users with 900 to 1100 followers since we believe that it is unlikely that a Swedish bot account has more than 1 million followers. In total, the two sets consist of 519 human accounts and 355 bot accounts.

The datasets used in [4, 10, 19] are not available in the original form. Either data is missing, or only the annotated labels of the accounts are given. Since it is not possible to obtain these datasets in their original form, we cannot use the datasets for comparing performance. The datasets were only used for training our model.

The dataset used in [6] (referred to as test set one by the authors of the paper) is the only dataset available in its original form. The dataset consists of 991 social spam bots and 991 genuine accounts. The genuine accounts are randomly selected from a set of more than 3000 accounts to get a 50/50 distribution of bots and genuine accounts. This means that we do not have the same set as the authors. The bots are collected in conjunction with a mayoral election in Rome 2014 where one candidate bought 1000 automatic accounts. The purchased accounts all had (stolen) profile pictures, (fake) profile description and a (fake) location. Genuine accounts were identified by sending out a question to randomly selected Twitter users. The ones that replied were considered genuine.

Features

In our classification model, we have used a total of 140 features. The features can be divided into two different types. The first type is *User Meta Data features* where information about the characteristics of the profile, such as the number of followers and friends and a total number of tweets is gathered. The second feature type is the *tweet features* that holds information about the actual content and when and how the content is posted. Similar to what is done in [2, 19] we use text entropy assuming that bots might have a less complex and varied way of expressing themselves. Similar to [19], we have included time features such as statistics of time between consecutive tweets, retweets, and mentions as well as statistics for the time between posted tweets containing URLs. All features are listed in Table 1.

Classification Algorithm

There have been several approaches to build classification models for bot detection. Different algorithms such as AdaBoost, logistic regression, support vector machines and naive Bayes have been tested. The best results (so far) are when using random forest which is the motivation for us to also use random forest in our classification.

Model Evaluation

We have used three datasets from [4, 10, 19] together with our 140 different features to train a model using random forest. The model was tested on the only dataset we have access to in its original form. In [6], the same dataset was used to compare the performance of other bot classification models. We included a comparison from [6] and used the same dataset to test our model. The performance (Accuracy, Precision,

Table 1 List of the 140 features extracted from each Twitter user

Meta features	Content features
Age of account	# unique hashtags per tweet
# tweets	# unique mentions per tweet
# tweets per day	# unique Urls per tweet
Friends-account age ratio	Normalized distribution of sources
# followers	Time between tweets[a]
# friends	Length of tweet[a]
Follower-friends ratio	# unique sources
Has location	retweet-tweet ratio
Has default profile description	# hashtags per tweets
Has default profile image	# urls per tweet
# likes given	# mentions per tweet
# likes given per # followers	# media per tweet
# likes given per # friends	# symbols per tweet
# likes per day	# retweets achieved per # tweet
Length of user name	Time between urls[a]
	Time between mentions[a]
	Time between retweets[a]
	# words[a]
	Hours of day tweeting
	Weekdays tweeting
	Normalized distribution hours tweeting
	Normalized distribution weekdays tweeting
	Normalized distribution of tweet endings
	String entropy[a]
	Total entropy of all tweets' strings concatenated

[a] Statistics of an array of values (mean, median, population standard deviation, standard deviation, maximum value and minimum value)

Table 2 Performance measures for different models, the model with the best performance is marked with bold

Model	Type	A	P	R	F1
Our model	supervised	0.957	0.941	**0.976**	0.958
Davis et al.[7]	Supervised	0.734	0.471	0.208	0.288
Yang et al. [21]	Supervised	0.506	0.563	0.170	0.261
Miller et al. [16]	Unsupervised	0.526	0.555	0.358	0.435
Ahmed et al. [1]	Unsupervised	0.943	0.945	0.944	0.944
Cresci et al. [5]	Unsupervised	**0.976**	**0.982**	0.972	**0.977**

Recall, and F1-score) is shown in Table 2. The comparison includes both supervised and unsupervised models. As mentioned earlier, the 911 genuine accounts in [6] were selected randomly from a set of more than 3000 genuine accounts. Since our genuine accounts were selected randomly, we (most likely) ended up with a slightly different testing set.

The supervised methods use cluster algorithms to identify clusters of bots. İn [16] feature vectors with the majority of features as text features are clustered using DenStream and StremKM++ as clustering algorithms.

In [1] graph clustering on statistical features related to hashtags, URLs, mentions, and retweets are used. The feature vectors were compared to each other using Euclidean distance and then clustered using the Fast greedy community detection algorithm. In [5] a bio-inspired technique for modelling behavior of users online with so-called *digital DNA* sequences is presented. The sequences are string encodings of the behavior of a user, and the sequences are then compared between the different users by measuring the longest common substring to find clusters of users.

The results in Table 2 shows that the model from [5] performs best on all metrics except for recall where our model performs better. Our model performs best of the supervised models included in the comparison. We are aware that one of our training sets has the same author as our test set which might be a reason for the high accuracy that we obtain. However, we have verified that none of the users are found in both datasets.

Feature Importance

To get an understanding of what features that played an important role in the classification we have calculated the feature importance with *forests of trees*. The ten most important features are shown in Table 3.

The most important feature when determining whether an account is a bot or not is the number of given likes divided by the number of friends. The second most important feature is the ratio between the number of followers and friends. As mentioned earlier, this feature could however sometimes be misleading since bots can be able to unfollow accounts that not are following back. Several of the most important features are related to the time between retweets.

Table 3 Most important features for our bot classification

Top 10	Feature
1	# given likes per # friends
2	Followers-friends ratio
3	Maximum time between retweets
4	# retweets achieved per tweet
5	Standard deviation of time between retweets
6	Median time between retweets
7	Population standard deviation of time between retweets
8	Mean time between retweets
9	# given likes
10	# given likes per # followers

Table 4 The different categories used to categorize the content of tweets

Criticism om media	Criticism of journalists, journalism or media
Criticism of elites	Criticism of the government, or persons or organizations that are regarded as influential political elite
Party support	Support of one or more parties
Criticism of parties	Criticism of one or more parties
Criticism of immigration	Criticism of immigration, asylum seekers and refugees
Election fraud	Discussions or observations of election fraud
Deleted	Tweets that have been deleted and can no longer be analyzed
Other/Uncategorized	Content that does not fall into any of the categories above

2.2 Content Classification

To understand the difference in communicated messages between bots and genuine Twitter accounts, a random sample of tweets was selected and then manually coded. This resulted in a number of different categories. The categories were identified by five persons that independently analyzed a set of tweets. The categories were based on the content of the tweet. The result of the five persons' classifications was combined into eight generic categories that are described in Table 4.

When the categories were defined, ten research assistants from Uppsala University classified a total of 1063 tweets, 547 tweets were published by bots and 516 published by genuine account. Each tweet was classified by at least two research assistants and each tweets category was determined by a majority decision. The research assistants did not know whether the tweet they were classifying was published by a bot or a genuine account. If the category for a tweet was selected as party support or criticism of party, the research assistant was able to enter which one or which parties that the support or criticism was expressed for. If a tweet expressed both support and criticism, the research assistants were advised to select party support.

2.3 Data

The data that is analysed was collected during the period March 5th–September 30th 2018. All tweets in Swedish that including at least one of the hashtags #valet2018, #val2018 or #valet, or one of the keywords *valet2018*, *val2018* or *valet* (all words are related to the election). The word *valet* is *the election* in Swedish. A total of 1,005,276 tweets published by 70,973 accounts were collected from the Twitter streaming API.

The classification of the accounts in the dataset was done after all the tweets had been downloaded. A number of tweets in the dataset belong to accounts that were suspended by Twitter or deleted by the users themselves. The most common reason for an account to be suspended is that the account is spamming or using a fake identity.[1] This indicates that a large part of the suspended accounts are automated, however, accounts can also be suspended by exhibiting offensive behavior.[2]

3   The Amount of Bots

During the period from March 5th to September 30th 2018, a total of 1,005,276 tweets linked to the Swedish general election was published by 70,973 accounts. These accounts were classified into four different classes: genuine, automatic, suspended, or deleted. The distribution between the different classes of accounts are shown in Table 5. Most of the accounts were classified as genuine accounts.

Approximately 6% of the accounts that tweet about Swedish election were classified as automated accounts, in total they were responsible for 8% of the content. If we make the somewhat extreme assumption that all suspended accounts are automated as well, then 12% of accounts are automated, and 9% of the content is automated. The real proportion of automatic accounts are most likely somewhere in that interval.

In Fig. 1, the number of tweets per day is shown. From the beginning of August until the election day, there is an increase of tweets about the election. On the election day September 9th, more than 45 thousand tweets were published. The figure shows that the interest in discussing the election decreased right after the election.

In Fig. 2, the number of active accounts per month from March to September are shown. More accounts were participating in discussing the election closer to the election. The genuine accounts make up the biggest share of accounts for the whole period. From July to August, the number of genuine accounts increased with 118%. For the same period, July to August, the number of bots increased by 167%. From

Table 5 Distribution between the different categories of accounts

Account category	Number of accounts	Number of tweets
Genuine	60,384	876,792
Bots	4084	73,723
Suspended	4052	14,902
Deleted	2453	39,859

---

[1]See Twitter's help page about suspended accounts https://help.twitter.com/en/managing-your-account/suspended-twitter-accounts.

[2]See Twitter's end-user licence agreement https://help.twitter.com/sv/rules-and-policies/twitter-rules.

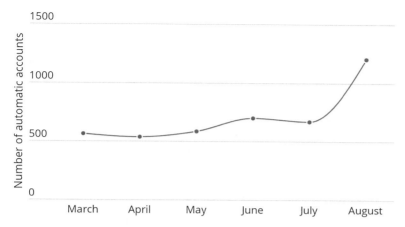

Fig. 1 The number of tweets per day for each category of accounts

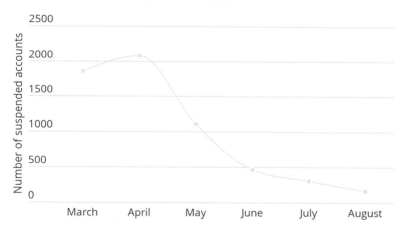

Fig. 2 The number of accounts per month for each category of accounts

August to September, the number of genuine accounts increased with 16%, and the numer of bot accounts with 28%.

4   Content

To get an understanding of what kind of content that were published we conducted two analyses. We explored the domains to which different accounts link to and what messages that are spread by the different classes of accounts.

4.1   Out Links

A tweet can consist of a maximum of 280 characters, which sometimes restricts the
user from actually convey the message in the text. It is quite common to include
a link to an external website in the tweet. In the dataset being analyzed here, it is
common to link to online versions of traditional media such as The Swedish public
service television company (SVT) or the newspaper Expressen, it is also common to
link to digital platforms such as Youtube and Facebook. A third category of media
that is quite common to link to is immigration critical alternative media[3] such as
Samhällsnytt and Nyheter idag.

To get an understanding regarding the differences between the different classes
of accounts, we have studied the most common domains that is linked to, for each
of the different classes of accounts. The ten most linked domains for each of the
categories of accounts are shown in Table 6. For each of the categories of accounts,
expressen.se and svt.se are the most common domains to link to. Youtube.com and
aftonbladet.se are found in the list of most linked domains for each of the class of
accounts. This is also true for the immigration critical alternative media samnytt.se,
nyheteridag.se, and friatider.se. For the suspended accounts, half of the ten most
linked domains are immigration critical. Here we can find the Nordic resistance
movement's news portal nordfront.se. One of the reasons for finding nordfront.se in
the list for the suspended accounts is that Twitter are suspending accounts which are
linking to nordfront.se.[4]

Table 7 shows the difference between the bots and genuine accounts most
shared domains 1 week before and after the election. Svt.se, expressen.se and

Table 6 The ten most linked domains for the different categories of accounts

Genuine	Bots	Suspended	Deleted
expressen.se	expressen.se	expressen.se	expressen.se
svt.se	svt.se	svt.se	svt.se
aftonbladet.se	aftonbladet.se	nordfront.se	samnytt.se
dn.se	omni.se	youtube.com	aftonbladet.se
youtube.com	sverigesradio.se	friatider.se	friatider.se
samnytt.se	youtube.com	samnytt.se	youtube.com
nyheteridag.se	samnytt.se	nyheteridag.se	nyheteridag.se
omni.se	friatider.se	aftonbladet.se	facebook.com
gp.se	dn.se	katerinamagasin.se	dn.se
friatider.se	nyheteridag.se	sverigesradio.se	gp.se

---

[3]In this chapter, we call these websites *immigration critical alternative media* since that is an
established concept in Sweden when talking about media that are critical of what is considered to
be an overly generous immigration policy.

[4]This article describes how Twitter removed accounts spreading links to nordfront.se (in Swedish)
https://www.dn.se/nyheter/politik/flera-konton-kopplade-till-nazistiska-nmr-raderade-av-twitter/.

Table 7 The ten most linked domains for bots and genuine accounts 1 week before and after the election

Genuine accounts		Bots	
Before election	After election	Before election	After election
svt.se	expressen.se	expressen.se	svt.se
expressen.se	svt.se	svt.se	expressen.se
aftonbladet.se	aftonbladet.se	aftonbladet.se	aftonbladet.se
nyheteridag.se	data.val.se	samnytt.se	youtube.com
samnytt.se	metro.se	sverigesradio.se	omni.se
youtube.com	dn.se	youtube.com	sverigesradio.se
dn.se	nyadagbladet.se	omni.se	dn.se
omni.se	omni.se	nyheteridag.se	samnytt.se
sverigesradio.se	youtube.com	friatider.se	metro.se
facebook.com	samnytt.se	dn.se	data.val.se

aftonbladet.se, all traditional media, are dominating as most shared domains both before and after the election. Links to the immigration critical domains have decreased after the election for both type of accounts. Data.val.se is a website with statistics about the election, it is present for both account classes after the election.

5 Messages

To understand the difference of content spread by bots and genuine accounts, we have analyzed a sample of tweets published 1 week before and 1 week after the election (see Sect. 2.2). The distribution between the different messages for bots and genuine accounts is shown in Fig. 3.

The most common message for both categories is party support followed by criticism of parties. More than 20% of the tweets published by bots have been deleted, compared to 10% for the genuine accounts. Overall, there are only minor differences between what bots and genuine accounts communicate.

In Figs. 4 and 5 the distribution of party support and party criticism for bots and genuine accounts are shown. The figures show the proportion of support and criticism for the different parties.

Regarding party support, it is clear that the Sweden democrats (SD) and Alternative for Sweden (AFS) are the most common parties to express support for in Twitter. It is also more common that a bot is showing support for SD compared to a genuine account. Also support for the Left Party (V) is more common from a bot than a genuine account.

Regarding the party criticism that is shown in Fig. 5, the criticism for the parties is relatively even between genuine accounts and bots. The Social Democratic Party (S) is the party receiving most of the criticism. The Center Party (C) and the Sweden democrats (SD) are the parties where the difference between criticism between

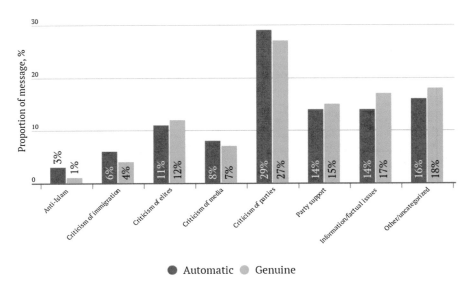

Fig. 3 The share of different messages spread by different categories of accounts

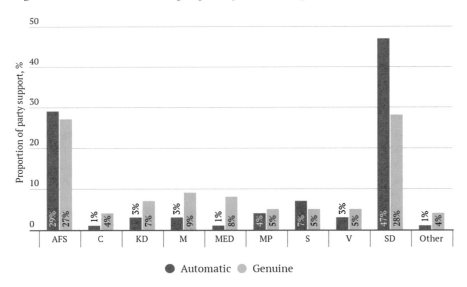

Fig. 4 The share of parties for which different categories of accounts express support

genuine and bot accounts are the largest. It is more common that a genuine account expresses criticism for SD, while it is more common for a bot to express criticism towards C.

There have been other studies on how bots were used in discussions about the Swedish election. In [3], the authors found at least 55 accounts with bot-like

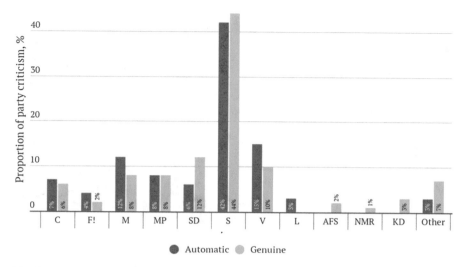

Fig. 5 Distribution of the parties for which different accounts express criticism

behavior that expressed support for AFS. Those accounts were active especially the weeks prior the election. The study showed that bot-like behavior was particularly clear for accounts criticizing S and supporting SD, which is in line with the results of our study. In [3] it is concluded that the behavior of the identified bot accounts was more genuine-like than bots which had been identified in previous elections in Germany and France.

The Swedish newspaper Dagens Nyheter published an article [15] about a Twitter analysis regarding the support of AFS. All tweets with the hashtag #afs18 was downloaded. The analysis showed that half of the 14,000 tweets with the hashtag was published by 15 accounts.

6   Spread

How messages spread and users communicate with each other can be understood by studying the network of accounts retweeting each other. Retweeting is a powerful tool for influencing others on Twitter. When retweeting another account, you are publishing someone else's post in your own feed. This is a way to distribute someone else's message and commonly this also means that you agree and express support for another account and the published post.

A principle in marketing and propaganda is that increased visibility leads to increased opportunity to influence. An article which because of several retweets appears in a user's feed several times will most likely influence the user more then an article that appears only once. If the article also conveys a message which

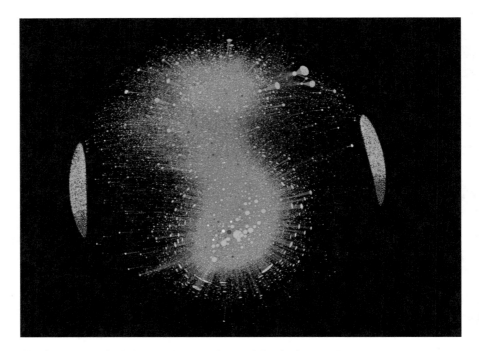

Fig. 6 The network of accounts that have discussed the election

the user recognizes from previous discussions, the article is even more likely to assimilate [8].

In Fig. 6, the 70,973 accounts tweeting about the election from March 5th until September 30th are shown. In the figure, every account is indicated by a node (circle), and every line between two accounts shows that one account has retweeted the material of another account in its own feed at least once during the time period. The size of an account's circle indicates how much an account has been retweeted by other users. The larger the circle, the more popular account to retweet. Green nodes represent genuine, red nodes represent bots, yellow nodes represent suspended, and the white nodes represent deleted accounts.

In Fig. 6, two big clusters of accounts which do not retweet anyone else in the networks are present. These two clusters appear in the top and bottom of the figure. Since these accounts have not retweeted anyone else in the network, they are not a part of the bigger cluster in the middle but are instead found outside the big cluster. This does not exclude that these users might follow, like and comment other users in the network. These accounts may also have ended up in our dataset since they have used the selected hashtags and keywords, but not in the context of discussions about the Swedish election.

Notably, there are a number of smaller clusters of accounts on the edges of the big cluster. These clusters consist of accounts that show the same behaviors as others in the groups, and they retweet only larger and more popular accounts. These

Fig. 7 The network of accounts that have discussed the election with the two clusters marked

small clusters indicate that some accounts have a group of followers which by only retweeting one account make sure that the account gets increased distribution of communicated messages. By investigating what category of accounts that appears in these smaller clusters, we can detect whether some accounts have used for example bots to get increased distribution of tweets.

Manual content analysis revealed that the main part of the network can be divided into two clusters. One of the clusters are inside the solid circle marked in Fig. 7. In this cluster, one can find the majority of the official accounts of the political parties in the Swedish parliament. The discussions are politically general and many different political issues are discussed. The second cluster can be found in the dashed circle. This cluster consists of accounts that discusses immigration policies and the negative consequences with immigration. The most popular accounts (in terms of retweets) in the network appear in the middle of the dashed cluster. These accounts have many followers and are often retweeted by several other accounts. In the middle between the two clusters (solid circled and dashed circled), we find accounts that positions themselves as political independent. These accounts are retweeted by both the clusters. Accounts that appears between the clusters can for example be news sites.

6.1   The Impact of Bots and Suspended Accounts

To get an understanding of whether the bots was effective in spreading messages, we investigated how the different account categories appears in the network. Since the most common reason that an account is being suspended by Twitter is that it shows what we call a bot-like behavior, we have also included the suspended accounts in this analysis.

The network analysis can be used to give answers to the following questions:

- Are the bots and the suspended accounts popular to retweet?
- Are there clusters consisting of bots and suspended accounts?

By detecting clusters of bots and suspended accounts, we can find networks connected to individual accounts that might have used bots to distribute their messages. Seven of the ten most retweeted accounts in the cluster are part of the immigration critical cluster. Nine of the accounts are genuine, and the tenth has been suspended by Twitter.

In Fig. 8, we can see where in the network the bots appear. The majority of the bots can be found outside the big cluster and appears in the two clusters of accounts

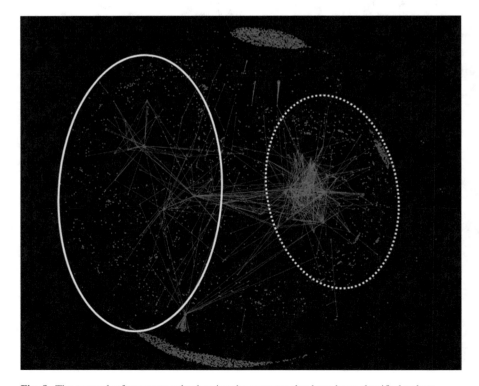

Fig. 8 The network of accounts only showing the accounts that have been classified as bots

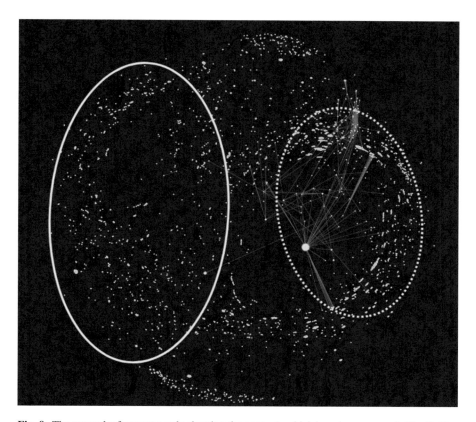

Fig. 9 The network of accounts only showing the accounts which have been suspended by Twitter

which has not retweeted nor been retweeted. The density of bots are higher in the immigration critical cluster but no large groups consisting of a large amount of bots retweeting only one account are identified.

In Fig. 9, we can see where in the network the accounts that have been suspended by Twitter appear. It is clear that the majority of the suspended accounts appear in the immigration critical cluster, and several smaller clusters of suspended accounts are found. Each of these clusters consist of accounts that have only retweeted one other account in the network and therefore acts as distributors of another account's messages. The occurrence of retweet accounts can serve as an indication that bots have been used.

Our analysis shows that it is more common for accounts discussing immigration to have other accounts distributing their messages. These accounts are more likely to be suspended by Twitter, one possible reason for the suspension is that they show a bot-like behavior.

7 Party Support on Twitter and the Election Results

In Sect. 5, we presented how bots and genuine accounts were expressing support for the different parties. In Fig. 10, the expressed party support on Twitter and the actual election results are shown. SD and AFS received the most support on Twitter while S and M got the largest share of votes in the election.

When the result of the election was reported, discussions about election fraud started on Twitter. The discussions included claims that the election was rigged and that the results were settled earlier and invalid. The term "valfusk" (election fraud in Swedish) occurred almost 2500 times the day after the election, as can be seen in Fig. 11.

The hashtag #valfusk was used frequently during the election night. Note that this hashtag was not a part of our searched hashtags or keywords and therefore all

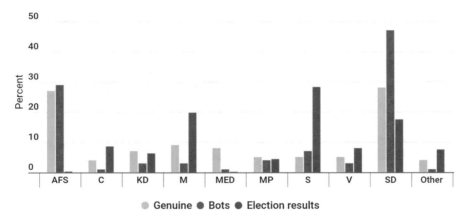

Fig. 10 Distribution of the parties for which different accounts express support, and the actual election results

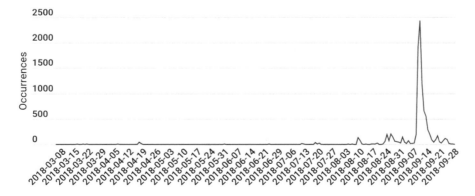

Fig. 11 Occurrences of the term valfusk (election fraud)

tweets with the hashtag #valfusk are not included in this study. The figure shows the occurrences of the term valfusk in the data with the hashtags and keywords mentioned in Sect. 2.

There are several reasons for the increase of discussion about election fraud. One reason might be many Twitter users were disappointed of the election results. Another reason can be that the experienced support of certain parties in Twitter did not agree with the results from the election.

8 Discussion and Conclusions

Among the accounts that we analyzed, the proportion of bots was between 6 and 12%, depending on whether we include accounts that were suspended among our bots or not. We can not say for sure how many of the suspended accounts that are bots, but since Twitter states that the most common reason for suspending an account is that the account shows bot-like behavior, we can assume that a high proportion of suspended accounts are bots. The number of bots that tweeted about the election more than doubled from July to August.

Our analyses shows that it has been more common for a bot account to express support for the Sweden democrats (SD), compared to a genuine account. The majority of the bots have not been a part of the cluster expressing immigration criticism, but rather a part of the clusters with accounts not retweeting any other accounts.

Our content analysis shows that criticism towards immigration and the negative consequences with immigration is frequently discussed on Twitter in conjunction with discussions about the election. However, it is important to point out that the hashtags we have used to gather our data might have been more popular to use among specific groups. This might have lead to skew in the result making the result in the study not a fully valid representation of the discussions about the election occurring on Twitter. The use of bots increased as the election date approached. The party AFS, which was founded in March 2018, did succeed on Twitter where they got around 25% of the party support from both bots and genuine accounts. This result is consistent with other studies that found that very active accounts with bot-like behavior has expressed support for the party.

It is important to add that it is not illegal to recruit bots to get an increased spread of a message. A thousand followers can be bought for around ten dollars[5] and retweets can be bought for twice as much.[6]

The analysis of supported parties showed that the immigration critical parties SD and AFS had more support on Twitter than in the actual election. After the election, several discussions about election fraud occurred. The difference between

---

[5]BuyTwitterFollowersReview—https://buytwitterfollowersreview.org/top-10/.
[6]BuySocialMediaMarketing—https://buysocialmediamarketing.com/twitter/retweets.

the election results and the expressed support on Twitter have likely contributed to the discussions about election fraud. Only a few days after the election, the discussions about election fraud subsided. One obvious conclusion of our study is that in digital media certain parts of the Swedish population are more engaged. In the case of Twitter, the most engaged are those who support the SD or who are strong critics of S.

One of the questions raised by this study concerns how bots influence the democratic process and the political discussion before the election. This question cannot be answered without an understanding of how the automatic accounts analyzed here fit into a larger picture of media use in Sweden. It would also be necessary to analyze who is behind these bots and what the aims are of spreading messages from bots. While answering these question is difficult, it is clear that the use of bots for spreading various types of messages increased the closer we come to the election. This could be a sign of an attempt to influence public opinion or at least a certain part of the political discussions.

An important question concerning all types of influence is to what extent the individual who is the target of influence is aware of whether someone is attempting to exercise influence. The results of previous research indicate that attempts to influence are less effective if individuals are aware of them [20]. In other words, an awareness that someone is trying to influence us can, at least to some extent, make us less susceptible to influence. Hopefully, studies like ours contributes to better awareness of attempts to exercise influence using bots.

References

1. F. Ahmed, M. Abulaish, A generic statistical approach for spam detection in online social networks. Comput. Commun. **36**(10-11), 1120–1129 (2013)
2. Z. Chu, S. Gianvecchio, H. Wang, S. Jajodia, Who is tweeting on twitter: human, bot, or cyborg? in *Proceedings of the 26th annual computer security applications conference* (ACM, New York, 2010), pp. 21–30
3. C. Colliver, P. Pmerantsev, A. Applebaum, J. Birdwell, Smearing Sweden, international influence campaigns in the 2018 Swedish election. Technical report (Institute of Strategic Dialogue (ISD), London School of Economics, London, 2018)
4. S. Cresci, R.D. Pietro, M. Petrocchi, A. Spognardi, M. Tesconi, Fame for sale: efficient detection of fake twitter followers. CoRR, abs/1509.04098 (2015)
5. S. Cresci, R.D. Pietro, M. Petrocchi, A. Spognardi, M. Tesconi, Dna-inspired online behavioral modeling and its application to spambot detection. IEEE Intell. Syst. **31**(5), 58–64 (2016)
6. S. Cresci, R.D. Pietro, M. Petrocchi, A. Spognardi, M. Tesconi, The paradigm-shift of social spambots: evidence, theories, and tools for the arms race (2017). CoRR, abs/1701.03017
7. C.A. Davis, O. Varol, E. Ferrara, A. Flammini, F. Menczer, Botornot: a system to evaluate social bots (2016). CoRR, abs/1602.00975
8. P.M. DeMarzo, D. Vayanos, J. Zwiebel, Persuasion bias, social influence, and unidimensional opinions. Q. J. Econ. **118**(3), 909–968 (2003)
9. R.S.J. Fernquist, L. Kaati, Political bots and the Swedish general election, in *International Conference of Security Informatics (ISI)* (2018)

10. Z. Gilani, R. Farahbakhsh, G. Tyson, L. Wang, J. Crowcroft, Of bots and humans (on twitter), in *Proceedings of the 2017 IEEE/ACM International Conference on Advances in Social Networks Analysis and Mining 2017*, ASONAM '17 (ACM, New York, 2017), pp. 349–354
11. R. Gorwa, D. Guilbeault, Understanding bots for policy and research: challenges, methods, and solutions (2018). CoRR, abs/1801.06863
12. M. Grusell, Sociala medier i svenska medierörelser, in *När makten står på spel—journalistik i valrörelser. Sthlm, Inst. f. mediestudier* (2017)
13. P.N. Howard, S. Woolley, R. Calo, Algorithms, bots, and political communication in the us 2016 election: the challenge of automated political communication for election law and administration. J. Inform. Tech. Polit. **15**(2), 81–93 (2018)
14. K. Lee, B. David Eoff, J. Caverlee, Seven months with the devils: A long-term study of content polluters on twitter, in *Fifth International AAAI Conference on Weblogs and Social Media*, vol. 01 (2011)
15. E. Mannheimer, H. Ewald, Så sprider falska konton högerextrema budskap inför eu-valet, in *Dagens nyheter* (2018)
16. Z. Miller, B. Dickinson, W. Deitrick, W. Hu, A.H. Wang, Twitter spammer detection using data stream clustering. Inf. Sci. **260**, 64–73 (2014)
17. N. Newman, R. Fletche, A. Kalogeropoulos, D.A.L. Levy, R.K. Nielsen, *Reuters Institute Digital News Report 2018* (Reuters Institute for the Study of Journalism/University of Oxford, Oxford, 2018)
18. M. Singh, D. Bansal, S. Sofat, Who is who on twitter–spammer, fake or compromised account? a tool to reveal true identity in real-time. Cybern. Syst. **49**(1), 1–25 (2018)
19. O. Varol, E. Ferrara, C.A. Davis, F. Menczer, A. Flammini, Online human-bot interactions: Detection, estimation, and characterization (2017). CoRR, abs/1703.03107
20. W. Wood, J.M. Quinn, Forewarned and forearmed? two meta-analysis syntheses of forewarnings of influence appeals. Psychol. Bull. **1**(129), 119–138 (2003)
21. C. Yang, R.C. Harkreader, G. Gu, Empirical evaluation and new design for fighting evolving twitter spammers. IEEE Trans. Inf. Forensics Secur. **8**(8), 1280–1293 (2013)

Cognitively-Inspired Inference for Malware Task Identification

Eric Nunes, Casey Buto, Paulo Shakarian, Christian Lebiere, Stefano Bennati, and Robert Thomson

Abstract Malware reverse-engineering, specifically, identifying the tasks a given piece of malware was designed to perform (e.g., logging keystrokes, recording video, establishing remote access) is a largely human-driven process that is a difficult and time-consuming operation. In this chapter, we present an automated method to identify malware tasks using two different approaches based on the ACT-R cognitive architecture, a popular implementation of a unified theory of cognition. Using three different malware collections, we explore various evaluations for each of an instance-based and rule-based model—including cases where the training data differs significantly from test; where the malware being evaluated employs packing to thwart analytical techniques; and conditions with sparse training data. We find that our approach based on cognitive inference consistently out-performs the current state-of-the art software for malware task identification as well as standard machine learning approaches—often achieving an unbiased F1 score of over 0.9.

1 Introduction

Identifying the tasks a given piece of malware was designed to perform (e.g. logging keystrokes, recording video, establishing remote access, etc.) is a difficult and time consuming task that is largely human-driven in practice [1]. The complexity of this task increases substantially when you consider that malware is constantly evolving, and that how each malware instance is classified may be different based on each

E. Nunes (✉) · C. Buto · P. Shakarian
Arizona State University, Tempe, AZ, USA
e-mail: enunes1@asu.edu; cbuto@asu.edu; shak@asu.edu

C. Lebiere · S. Bennati
Carnegie Mellon University, Pittsburgh, PA, USA
e-mail: cl@cmu.edu

R. Thomson
United States Military Academy, West Point, NY, USA
e-mail: robert.thomson@westpoint.edu

© Springer Nature Switzerland AG 2020
M. A. Tayebi et al. (eds.), *Open Source Intelligence and Cyber Crime*, Lecture Notes in Social Networks, https://doi.org/10.1007/978-3-030-41251-7_7

cyber-security expert's own particular background. Automated solutions for this problem are highly attractive as they can significantly reduce the time it takes to conduct remediation in the aftermath of a cyber-attack.

Earlier work has sought to classify malware by similar "families", something which has been explored as a supervised classification problem [2–4]. However, differences over determining "ground truth" for malware families (i.e. Symantec and McAfee cluster malware into families differently) and the tendency for automated approaches to only succeed at "easy to classify" samples [5, 6] are two primary drawbacks of malware family classification. More recently, there has been work on directly inferring the tasks a malware was designed to perform [7]. This approach leverages static malware analysis (i.e. analysis of the malware sample conducted without execution, such as decompilation) and a comparison with a crowd-source database of code snippets using a proprietary machine learning approach. However, a key shortcoming of the static method is that it is of limited value when the malware authors encrypt part of their code—as we saw with the infamous Gauss malware [8]. This work builds upon recent developments in the application of cognitive models to intelligence analysis tasks [9] and our own preliminary studies on applying cognitive models to identify the tasks a piece of malware was designed to perform [10, 11]. Specifically, in this chapter, we report

- Experimental results illustrating consistent and significant performance improvements (in terms of precision, recall, and F1) of the instance-based cognitive model approach when compared with various standard machine learning approaches (including SVM, logistic regression and random forests) for two different sandboxes and for three different datasets.
- Experimental results showing a consistent and significant performance improvement of the instance-based cognitive model and several other machine learning approaches when compared to the current state-of-the-art commercial technology (based on static analysis).
- Experiments where we study cases where the malware samples are mutated, encrypted, and use different carriers—providing key insights into how our approach will cope with operational difficulties.
- Experimental results illustrating that a cognitively-inspired intermediate step of inferring probability distribution over malware families provides improved performance over the machine learning and rule-based cognitive model (though no significant change to the instance-based cognitive model).

Cognitive models have proved to significantly outperform classical machine learning approaches and state of the art products available in the market for malware task prediction [10–12]. This chapter consolidates the results presented in previous research by including additional results for the GVDG dataset, runtime comparisons of the experiments and discussing the cognitive models in terms of parameter selection and time complexity analyses. We also provide new experimental results utilizing a dataset based on the MetaSploit framework [13]— demonstrating how our framework adapts to features based on malware that utilizes the network protocol stack. We also explore the concept of predicting hacker intentions on a host machine.

This chapter is organized as follows. In Sect. 2 we state the technical preliminaries used in the chapter. In Sect. 3.1 we introduce our cognitive-based approaches, describing the algorithms and explaining our selection of parameter settings. This is followed by a description of the baseline approaches that we studied in our evaluation in Sect. 4.1 and a description of the two different dynamic malware sandbox environments we used in Sect. 4.2. In Sect. 5 we present our suite of experimental results which include experiments involving samples discovered by Mandiant, Inc. in their APT1 report [14], samples created using the GVDG [15] tool, and samples created using the MetaSploit framework. Finally, related work is discussed in Sect. 6.

2   Technical Preliminaries

Throughout this chapter, we shall assume that we have a set of malware samples that comprise a historical corpus (which we shall denote $\mathcal{M}$) and each sample $i \in \mathcal{M}$ is associated with a set of tasks (denoted $tasks(i)$) and a set of attributes (denoted $attribs(i)$). Attributes are essentially binary features associated with a piece of malware that we can observe using dynamic and/or static analysis while the tasks—which tell us the higher-level purpose of the malware—must be determined by a human reviewing the results of such analysis. As $\mathcal{M}$ comprises our historical knowledge, we assume that for each $i \in \mathcal{M}$ both $tasks(i)$ and $attribs(i)$ are known. For a new piece of malware, we assume that we only know the attributes. We also note that throughout the chapter, we will use the notation $| \cdot |$ to denote the size of a given set. Tables 1 and 2 provide examples of the attributes and tasks based on the malware samples from the Mandiant APT1 dataset (created from samples available at [16], see also [14]). For instance, *hasDynAttrib* looks at the behavior section of the analysis report and extracts all the activity of the malware on the host machine. The attribute *usesDll* enumerates all the libraries that were used by the malware on the host machine. The file activity and the registry activity is captured by *fileAct* and *regAct*. Finally all the processes initiated and terminated by the malware are captured by *proAct*. There is not a fixed number of any of these attributes for a

Table 1 Attributes extracted through automated malware analysis

Attribute	Intuition
usesDll(X)	Malware uses a library X
regAct(K)	Malware conducts an activity in the registry, modifying key K.
fileAct(X)	Malware conducts an activity on certain file X
proAct	Malware initiates or terminates a process

Table 2 Sample of malware tasks

Task	Intuition
beacon	Beacons back to the adversary's system
enumFiles	Designed to enumerate files on the target
serviceManip	Manipulates services running on the target
takeScreenShots	Takes screen shots
upload	Designed to upload files from the target

given malware. The number of attributes depends on the analysis report generated from the sandbox. A full description of this dataset is presented in Sect. 5.

Throughout the chapter, we will also often consider malware families, using the symbol $\mathcal{F}$ to denote the set of all families. Each malware sample will belong to exactly one malware family, and all malware samples belonging to a given family will have the same set of tasks. Hence, we shall also treat each element of $\mathcal{F}$ as a subset of $\mathcal{M}$.

3 Cognitively-Inspired Inference

While human inference has memory and attentional limitations, their cognitive processes are powerful, where adaptive heuristic strategies are adopted to accomplish the tasks under strong time constraints using limited means. An advantage of using a cognitive model to describe inferential processes is that the underling architecture provides the benefits of human-inspired inference while allowing for more flexibility over constraints such as human working memory. We believe that there is a valid use of cognitive architectures for artificial intelligence that makes use of basic cognitive mechanisms while not necessarily making use of all constraints of the architecture. In that case, it is arguably better to specifically state which aspects of the model are not constrained by data, and rather than mock up those aspects in plausible but impossible to validate manner, simply treat them as unmodeled processes. This approach results in simpler models with a clear link between mechanisms used and results accounted for, rather than being obscured by complex but irrelevant machinery. For instance, while the models described in this chapter use activation dynamics well-justified against human behavioral and neural data to account for features such as temporal discounting, we do not directly model working memory constraints to allow for more features of malware and more instances to be present in memory.

3.1 ACT-R Based Approaches

We propose two models built using the mechanisms of the ACT-R (Adaptive Control of Thought-Rational) cognitive architecture [17]. These models leverage the work on applying this architecture to intelligence analysis problems [9]. In particular, we look to leverage our recently-introduced instance-based (ACTR-IB) and rule-based (ACTR-R) models [10, 11]. Previous research has argued the ability of instance-based learning in complex dynamic situations making it appropriate for sensemaking [18]. On the other hand, the rule-based learning is a more compact representation of associating samples in memory with their respective families. In this section, we review some of the major concepts of the ACT-R framework that are relevant to these models and provide a description of both approaches.

We leveraged features of the declarative memory and production system of the ACT-R architecture to complete malware task identification. In ACT-R, recall from declarative memory (c.f., identification, for our purposes) depends on three main components: activation strengthening (i.e., the base-level activation of an element), associative (i.e., spreading) activation, and inter-element similarity (i.e., partial matching). These three values are summed together to represent an item's total activation. When a recall is requested from memory, the item with the highest total activation is retrieved.

Declarative Knowledge Declarative knowledge is represented formally in terms of *chunks*. Chunks have an explicit type, and consist of an ordered list of slot-value pairs of information. Chunks are retrieved from declarative memory by an activation process, and chunks are each associated with an *activation strength* which in turn is used to compute a *retrieval probability*. In this chapter, chunks will typically correspond to a malware family. In the version of ACTR-IB where we do not represent families explicitly, the chunks correspond with samples in the training data.

For a given chunk i, the activation strength A_i is computed as,

$$A_i = B_i + S_i + P_i \tag{1}$$

where, B_i is the base-level activation, S_i is the spreading activation, and P_i is the partial matching score. We describe each of these in more detail as follows.

Base-Level Activation (B_i) Technically, base-level for chunk i reflects both the frequency and recency of samples in memory, even though we are not using recency here but it could easily be applicable to weigh samples toward the more recent ones. More important, base-level is set to the *log* of the prior probability (i.e., the fraction of samples associated with the chunk) in ACTR-R; for instance-based (ACTR-IB), we set it to a base level constant β_i.

Spreading Activation (S_i) Spreading activation is a measure of the uniqueness of the attributes between a test sample i and a sample j in memory. The spread of activation to sample i is computed by the summing the strengths of association between sample j and the attributes of the current sample i being considered. To compute the spreading activation we compute the *fan* of attribute a (i.e., the number of samples in memory with attribute a) for each attribute. The strength of association is computed differently in both approaches and, in some cognitive model implementations, is weighted (as is done in ACTR-R of this chapter).

Partial Matching (P_i) A partial matching mechanism computes the similarity between two samples. In this work, it is only relevant to the instance-based approach. Given a test sample j, its similarity with a sample i in memory is computed as a product of the mismatch penalty (mp, a parameter of the system) and the degree of mismatch M_{ji}. We define the value of M_{ji} to be between 0 and -1; 0 indicates complete match while -1 complete mismatch.

As common with models based on the ACT-R framework, we shall discard chunks whose activation strength is below a certain threshold (denoted τ). Once the activation strength, A_i, is computed for a given chunk, we can then calculate the activation probability, p_i. This is the probability that the cognitive model will recall that chunk and is computed using the Boltzmann (softmax) equation [19], which we provide below.

$$Pr_i = \frac{(e^{\frac{A_i}{s}})}{\sum_j (e^{\frac{A_j}{s}})} \tag{2}$$

Here, e is the base of the natural logarithm and s is momentary noise inducing stochasticity by simulating background neural activation (this is also a parameter of the system).

3.2   ACT-R Instance-Based Model

The instance based model is an iterative learning method that reflects the cognitive process of accumulating experiences (in this case the knowledge base of training samples) and using them to predict the tasks for unseen test samples. Each malware instance associates a set of attributes of that malware with its family. When a new malware sample is encountered, the activation strength of that sample with each sample in memory is computed using Eq. 1. The spreading activation is a measure of the uniqueness of the attributes between a test sample i and a sample j in memory. To compute the spreading activation we compute the fan for each attribute a ($fan(a)$ finds all instances in memory with the attribute a) of the test sample i. The Partial matching is computed as explained above. The degree of mismatch is computed as the intersection between the attribute vector of the given malware and each sample in memory normalized using the Euclidean distance between the two vectors. The retrieval probability of each sample j in memory with respect to the test sample i is then computed using Eq. 2. This generates a probability distribution over families. The tasks are then determined by summing up the probability of the families associated with that task with an appropriately set threshold (we set that threshold at 0.5 (indicates that the model should be more than 50% confident before a task is predicted for a test malware sample)). Algorithm 1 shows the pseudo code for the instance-based model.

Algorithm 1: ACT-R instance-based learning

INPUT: New malware sample i, historical malware corpus $\mathcal{M}$.
OUTPUT: Set of tasks associated with sample i.
for query malware sample i **do**
    for all j in $\mathcal{M}$ **do**
        $B_j = \beta_j$
        $P_j = mp \times \frac{|attribs(i) \cap attribs(j)|}{\sqrt{|attribs(i)| \times |attribs(j)|}}$
        for $a \in attribs(i)$ **do**
            if $a \in attribs(j)$ **then**
                $s_{ij} \mathrel{+}= log(\frac{|\mathcal{M}|}{|fan(a)|})$
            else
                $s_{ij} \mathrel{+}= log(\frac{1}{|\mathcal{M}|})$
            end if
        end for
        $S_j = \sum_j \frac{s_{ij}}{|attribs(i)|}$
        Calculate A_j as per Equation 1
    end for
    Calculate p_j as per Equation 2
    $p_f = \sum_{j \in f s.t. A_j \geq \tau} p_j$
    $t_p = \{t \in T | p_f \geq 0.5\}$
end for

Time Complexity of Instance-Based Model The Instance based model has no explicit training phase, so there are no training costs associated with it. For a given test sample the model computes the activation function for each sample in the knowledge base. Hence the time complexity increases linearly with the knowledge base. Let n be the number of the samples in the knowledge base and m is the number of attributes associated with the test sample, then the time complexity can be given as $O(nm)$ for each test sample, as we expect m to be relative small ($n >> m$), the relationship is linear in n.

3.3 ACT-R Rule-Based Model

In this version of ACT-R model we classify the samples based on simple rules computed during the training phase. Given a malware training sample with its set of attributes a, along with the ground truth family value, we compute a pair of conditional probabilities $p(a|f)$ and $p(a|\neg f)$ for an attribute in a piece of malware belonging (or not belonging) to family f. These probabilistic rules (conditional probabilities) are used to set the strength of association of the attribute with a family $(s_{a,f})$. The strength of association is weighted by the source activation w to avoid retrieval failures for rule-based models. We use empirically determined Bayesian priors $p(f)$ to set the base-level of each family as opposed to using a constant base-level for instance based. Only two components of the activation Equation 1 are used, namely the base-level and the spreading activation. Given the attributes for current malware, we calculate the probability of the sample belonging to each family according to Eq. 2, generating a probability distribution over families. The task are then determined in a similar way to that of instance-based model. Algorithm 2 shows the pseudo code for the rule-based model.

Algorithm 2: ACT-R rule-based learning

INPUT: New malware sample i, historical malware corpus $\mathcal{M}$.
OUTPUT: Set of tasks associated with new sample i.
TRAINING:
Let $X = \bigcup_{j \in \mathcal{M}} attrib(j)$
for all a in X **do**
    Compute the set of rules $p(a|f)$ and $p(a|\neg f)$
    (where $p(a|f) = \frac{|\{i \in \mathcal{M} \cap f | a \in attrib(i)\}|}{|f|}$
    and $p(a|\neg f) = \frac{|\{i \in \mathcal{M} - f | a \in attrib(i)\}|}{|\mathcal{M}| - |f|}$)
end for
TESTING:
for all $f \in \mathcal{F}$ **do**
    $B_f = log(p(f))$ (where $p(f) = \frac{|f|}{|\mathcal{M}|}$)
    for all $a \in attrib(i)$ **do**
        $s_{a,f} = log(\frac{p(a|f)}{p(a|\neg f)})$; $S_f =+ \frac{w \times s_{a,f}}{|attribs(i)|}$
    end for
    $A_f = B_f + S_f$
end for
Calculate p_f as per Equation 2
$t_p = \{t \in T | p_f \geq 0.5\}$

Time Complexity of Rule-Based Model For Rule-based model computing the rules for each attribute in the knowledge base significantly add to the computation time. Let n be the number of samples in the training set, m be the number of attributes in the new piece of malware, and m^* be the cardinality of $\bigcup_{j \in \mathcal{M}} attrib(j)$. The resulting time complexity for training is then $O(m^*n)$ for training, which is significant as we observed $m^* >> m$ in our study. While this is expensive, we note that for testing an individual malware sample, the time complexity is less than the testing phase for the instance based $O(|\mathcal{F}|m)$—though the instance based model requires no explicit training phase (which dominates the time complexity of the training phase for the rule-based approach).

Table 3 Parameters for the cognitive models

Model	Parameters
Instance based learning	$\beta = 20$ (base-level constant)
	$s = 0.1$ (stochastic noise parameter)
	$\tau = -10$ (activation threshold)
	$mp = 20$ (mismatch penalty)
Rule based learning	$s = 0.1$ (stochastic noise parameter)
	$w = 16$ (source activation)

3.4 Model Parameter Settings

The two proposed models leverage separate components of the activation function. Table 3 provides a list of parameters used for both the ACT-R models—we use standard ACT-R parameters that have been estimated from a wide range of previous ACT-R modeling studies from other domains [20] and which are also suggested in the ACT-R reference manual [21].

The intuition behind these parameters is as follows. The parameter s injects stochastic noise in the model. It is used to compute the variance of the noise distribution and to compute the retrieval probability of each sample in memory. The mismatch penalty parameter mp is an architectural parameter that is constant across samples, but it multiplies the similarity between the test sample and the samples in knowledge base. Thus, with a large value it penalizes the mismatch samples more. It typically trades off against the value of the noise s in a signal-to-noise ratio manner: larger values of mp lead to more consistent retrieval of the closest matching sample whereas larger values of s leads to more common retrieval of poorer matching samples. The activation threshold τ determines which samples will be retrieved from memory to make task prediction decisions. The base level constant β is used to avoid retrieval failures which might be caused due to high activation threshold. The source activation w is assigned to each retrieval to avoid retrieval failures for rule-based models.

4 Experimental Setup

4.1 Baseline Approaches

We compare the proposed cognitive models against a variety of baseline approaches—one commercial package and five standard machine learning techniques. For the machine learning techniques, we generate a probability distribution over families and return the set of tasks associated with a probability of 0.5 or greater while the commercial software was used as intended by the manufacturer. Parameters for all baseline approaches were set in a manner to provide the best performance.

Commercial Offering: Invencia Cynomix Cynomix is a malware analysis tool made available to researchers by Invencia industries [7] originally developed under DARPA's Cyber Genome project. It represents the current state-of-the-art in the field of malware capability detection. Cynomix conducts static analysis of the malware sample and uses a proprietary algorithm to compare it to crowd-sourced identified malware components where the functionality is known.

Decision Tree (DT) Decision tree is a hierarchical recursive partitioning algorithm. We build the decision tree by finding the best split attribute i.e. the attribute that maximizes the information gain at each split of a node. In order to avoid over-fitting, the terminating criteria was set to less than 5% of total samples. Malware samples are tested by the presence and absence of the best split attribute at each level in the tree till it reaches the leaf node. When it reaches the leaf node the probability distribution at the leaf node is assigned to the malware sample.

Naive Bayes Classifier (NB) Naive Bayes is a probabilistic classifier which uses Bayes theorem with independent attribute assumption. During training we compute the conditional probabilities of a given attribute belonging to a particular family. We also compute the prior probabilities for each family; i.e., fraction of the training data belonging to each family. Naive Bayes assumes that the attributes are statistically independent hence the likelihood for a sample S represented with a set of attributes a associated with a family f is given as, $p(f|S) = P(f) \times \prod_{i=1}^{d} p(a_i|f)$.

Random Forest (RF) Ensemble methods are popular classification tools. They are based on the idea of generating multiple predictors used in combination to classify new unseen samples. We use a random forest which combines bagging for each tree with random feature selection at each node to split the data, thus generating multiple decision tree classifiers [22]. Each decision tree gives its own opinion on test sample classification which is then merged to generate a probability distribution over families. For all the experiments we set the number of trees to be 100, which gives us the best performance.

Support Vector Machine (SVM) Support vector machines (SVM) are proposed by Vapnik [23]. SVMs work by finding a separating margin that maximizes the geometric distance between classes. The separating margin is termed as hyperplane. We use the popular LibSVM implementation [24] which is publicly available. The implementation has the option of returning the probability distribution as opposed to the maximum probability prediction.

Logistic Regression (LOG-REG) Logistic regression classifies samples by computing the odds ratio. The odds ratio gives the strength of association between the attributes and the family like simple rules used in the ACT-R rule based learning. We implement the multinomial logistic regression which handles multi-class classification.

4.2 Dynamic Malware Analysis

Dynamic analysis studies a malicious program as it executes on the host machine. It uses tools like debuggers, function call tracers, machine emulators, logic analyzers, and network sniffers to capture the behavior of the program. We use two publicly available malware analysis tools to generate attributes for each malware sample. These tools make use of a sandbox, which is a controlled environment to run malicious software.

Anubis Sandbox Anubis [25] is an online sandbox which generates an XML formatted report for a malware execution in a remote environment. It generates detailed static analysis of the malware but provides less details regarding the behavior of the malware on the host machine. Since it is hosted remotely we cannot modify its settings.

Cuckoo Sandbox Cuckoo [26] is a standalone sandbox implemented using a dedicated virtual machine and more importantly can be customized to suit our needs. It generates detailed reports for both static as well as behavior analyses by watching and logging the malware while its running on the virtual machine. These behavior analyses prove to be unique indicators (behavior patterns common to a single family) for a given malware for the experiments.

4.3 Performance Evaluation

In our tests, we evaluate performance based primarily on four metrics: precision, recall, unbiased F1, and family prediction accuracy. For a given malware sample being tested, precision is the fraction of tasks the algorithm associated with the malware that were actual tasks in the ground truth. Recall, for a piece of malware, is the fraction of ground truth tasks identified by the algorithm. The unbiased F1 is the harmonic mean of precision and recall. In our results, we report the averages for precision, recall, and unbiased F1 for the number of trials performed. Our measure of family accuracy—the fraction of trials where the most probable family was the ground truth family of the malware in question—is meant to give some insight into how the algorithm performs in the intermediate steps.

5 Results

All experiments were run on Intel core-i7 operating at 3.2 GHz with 16 GB RAM. Only one core was used for experiments. Except where explicitly noted, the ACT-R parameters were fixed as per Table 3 for all experiments (across all datasets and sandboxes).

5.1 *Mandiant Dataset*

Our first set of experiments uses a dataset based on the T1 cyber espionage group as identified in the popular report by Mandiant Inc [14]. This dataset consisted of 132 real malware samples associated with the Mandiant report that were obtained from the Contagio security professional website [16]. Each malware sample belonged to one of 15 families including BISCUIT, NEWSREELS, GREENCAT and COOK-IEBAG. Based on the malware family description [14], we associated a set of tasks with each malware family (that each malware in that family was designed to perform). In total, 30 malware tasks were identified for the given malware samples (Table 2). On average, each family performed nine tasks.

We compared the four machine learning approaches with the rule-based and instance-based ACT-R models (ACTR-R and ACTR-IB respectively). We also submitted the samples to the Cynomix tool for automatic detection of capabilities. These detected capabilities were then manually mapped to the tasks from the Mandiant report. Precision and recall values were computed for the inferred adversarial tasks. On average the machine learning approaches predicted nine tasks per sample, ACTR-R predicted nine tasks per sample and ACTR-IB predicted ten tasks. On the other hand, Cynomix was able to detect on average only four tasks.

Leave One Out Cross-Validation (LOOCV)
In leave one out cross validation, for n malware samples, train on $n - 1$ samples and test on the remaining one. This procedure was repeated for all samples and the results were averaged. We performed this experiment using both sandboxes and compared the results (Table 4).

The average F1 increases by 0.03 when we use the attributes generated by the Cuckoo sandbox instead of Anubis. The statistical significance results are as follows: for ACTR-IB (t (132) = 1.94, $p = 0.05$), ACTR-R (t (132) = 1.39, $p = 0.16$), RF (t (132) = 0.56, $p = 0.57$), SVM (t (132) = 1.95, $p = 0.05$), LOG-REG (t (132) = 1.82, $p = 0.07$), NB (t (132) = 1.79, $p = 0.08$) and DT (t (132) = 0.83, $p = 0.4$). But the significant improvement was in the family prediction values with ACTR-IB improving by 0.12 from 0.81 to 0.93 (t (132) = 3.86, $p < 0.001$) and ACTR-R by 0.15 from

Table 4 Performance comparison of Anubis and Cuckoo Sandbox (Bold values indicates best performance)

Method	Anubis (F1)	Cuckoo (F1)	Anubis (Family)	Cuckoo (Family)
DT	0.80	0.80	0.59	0.63
NB	0.71	0.74	0.30	0.40
LOG-REG	0.82	0.85	0.65	0.84
SVM	0.86	0.90	**0.85**	0.86
RF	0.89	0.89	0.82	0.86
ACTR-R	0.85	0.88	0.73	0.89
ACTR-IB	**0.93**	**0.96**	0.81	**0.93**

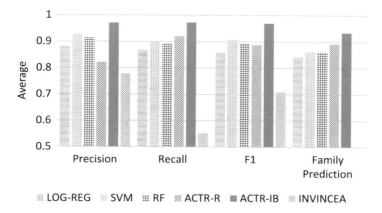

Fig. 1 Average precision, recall, F1 and family prediction comparisons using cuckoo sandbox for LOG-REG, RF, SVM, ACTR-R, ACTR-IB and INVINCEA

0.72 to 0.87 (t (132) = 3.78, $p < 0.001$) outperforming all other methods. Since having behavior analysis helps in better task prediction as seen from the comparison experiment, we use cuckoo sandbox for rest of our experiments.

Figure 1 compares the performance of the five best performing methods from Table 1 and compares it with the Cynomix tool of Invincea industries. ACTR-IB outperformed LOG-REG, SVM, RF and ACTR-R; average F1 = 0.97 vs 0.85 (t (132) = 7.85, $p < 0.001$), 0.9 (t (132) = 4.7, $p < 0.001$), 0.89 (t (132) = 5.45, $p < 0.001$) and 0.88 (t (132) = 5.2, $p < 0.001$) respectively. Both the proposed cognitive models and machine learning techniques significantly outperformed the Cynomix tool in detecting the capabilities (tasks).

These three approaches (LOG-REG, SVM, RF) were also evaluated with respect to predicting the correct family (before the tasks were determined). ACTR-IB outperformed LOG-REG, SVM, RF and ACTR-R; average family prediction = 0.93 vs 0.84 (t (132) = 3.22, $p < 0.001$), 0.86 (t (132) = 3.13, $p < 0.001$), 0.86 (t (132) = 3.13, $p < 0.001$) and 0.89 (t (132) = 2.13, $p = 0.03$) respectively. The Cynomix tool from Invincea does not have the capability to predict families.

Task Prediction Without Inferring Families

In the proposed models we infer the malware family first and then predict the tasks associated with that family. However, differences over "ground truth" for malware families in the cyber-security community calls for a direct inference of tasks without dependence on family prediction. In this section we adapt the models to predict tasks directly without inferring the family.

Figure 2 shows the performance of the cognitive and machine learning models without inferring the families. There is no difference in the performance of ACTR-IB and ACTR-R approaches as compared to Fig. 2 where we use families. On the other hand, direct task prediction reduces the F1 measure of machine learning techniques on average by almost 0.1. This is due to the fact that, now instead

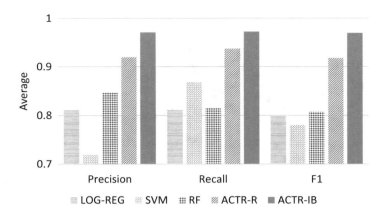

Fig. 2 Average precision, recall, and F1 comparisons for LOG-REG, RF, SVM, ACTR-R and ACTR-IB for Mandiant without inferring families

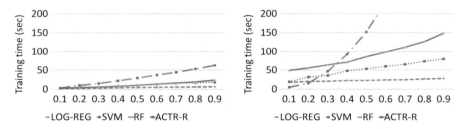

Fig. 3 Training time for LOG-REG, SVM, RF and ACTR-R with(left)/without(right) inferring families

of having a single classifier for each family we have multiple classifiers for each task that a malware sample is designed to perform. This not only degrades the performance but also adds to the training time for these methods (including the ACT-R rule-based approach). We compare the training time with increase in training data for task prediction with/without inferring families. Inferring families first reduces the training time (Fig. 3 (left)). On the other hand, predicting tasks directly significantly increases the training time for the machine learning methods along with the rule-based ACT-R approach (Fig. 3 (right)). Due to the issues with respect to performance and training time, we consider inferring families first for the rest of the experiments. An important point to note is that this has no effect on the Instance-based model for both performance and computation time.

Parameter Exploration

We now discuss two system parameters that control the performance of the ACT-R instance based model namely the stochastic noise parameter (s) and the activation threshold (τ). We use the mandiant dataset to perform this evaluation. The parameter s takes values between 0.1 and 1 (typical values range from 0.1 to 0.3). The value

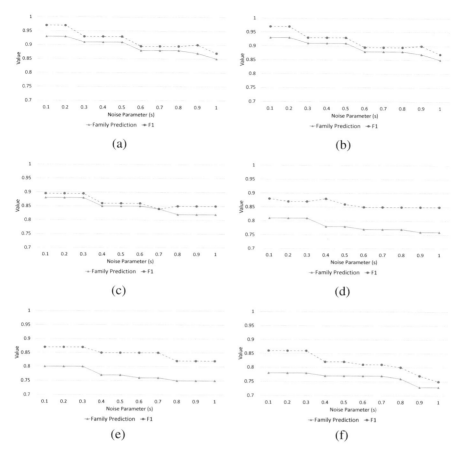

Fig. 4 Family prediction and F1 value for different threshold and noise parameters values. (**a**) $\tau = -20$. (**b**) $\tau = -10$. (**c**) $\tau = 0$. (**d**) $\tau = 5$. (**e**) $\tau = 10$. (**f**) $\tau = 15$

of the activation threshold depends on the application. Figure 4 shows the variation of family prediction accuracy and F1 score with respect to different noise parameter values and for different activation thresholds. The parameter s is used to compute the variance of the noise distribution and retrieval probability of sample in memory. Larger value of s triggers the retrieval of poor matching samples, which leads to lower family prediction and F1 scores. As seen in Fig. 4, as the value of s increases the performance decreases. On the other hand, the activation threshold dictates how many closely matched samples will be retrieved from memory. For high values of τ the performance decreases as many fewer samples are retrieved. For lower values of τ we end up retrieving almost all the samples in the training data, hence the performance does not decrease as τ decreases, but it adds to the computational cost of retrieving high number of samples which is not desirable. We get the best performance for $\tau = -10$ and $s = 0.1$. Even $s = 0.2$ is almost as good as 0.1 providing some advantages in terms of stochasticity ensuring robustness.

We keep the base-level constant (β) and mismatch penalty (mp) values constant. As explained earlier the base-level constant trades off directly against the retrieval threshold, and the mismatch penalty against the activation noise, respectively, so it makes sense to vary only one of the pair.

5.2   GVDG Dataset

GVDG is a malware generation tool designed for the study of computer threats [15]. It is capable of generating the following malware threats:

- File-virus
- Key-Logger
- Trojan-Extortionist
- USB-Worm
- Web Money-Trojan

Figure 5 shows the GVDG user interface used for the generation of malware samples. We can select the carrier type and the tasks that we want the malware sample to perform on the host machine. The tasks are represented as payloads, while carrier is a functional template which can be modified to execute the tasks desired by the user on the host system. In generating datasets with GVDG, we specify

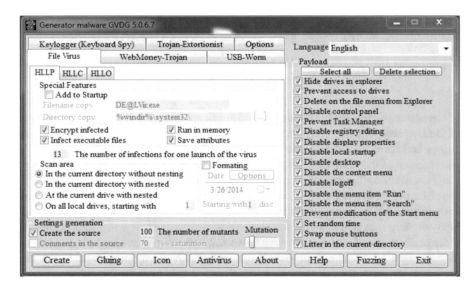

Fig. 5 GVDG user interface

families based on sets of malware with the same tasks. Whether or not a family consists of malware with the same carrier depends on the experiment. Further, GVDG also has an option to increase "mutation" or variance among the samples. We perform experiments analyzing the performance of the proposed methods when the generated samples belong to different carrier and same carrier types, as well as when the samples are encrypted and mutated making task prediction difficult. In all the experiments we consider 60% of the data for training and 40% for testing. The results are averaged across ten trials. The Cynomix tool from Invencia was unable to detect any tasks for the GVDG dataset, primarily due to its inability to find public source documents referencing GVDG samples and also unable to generalize from similar samples.

Different Carriers

In this experiment, we generated 1000 samples for each carrier type with low mutation. On average each carrier type performs seven tasks (payloads). Hence each carrier represents one family for this experiment. Both random forest and ACTR-IB model were able to predict the tasks and family with F1 measure of 1.0 outperforming LOG-REG 1 vs 0.91, SVM 1 vs 0.95 and ACTR-R 1 vs 0.95. All results are statistical significant with (t $(1998) \geq 8.93$, $p < 0.001$) (Fig. 6). Also for family prediction ACTR-IB and RF outperformed LOG-REG 1 vs 0.92, SVM 1 vs 0.92 and ACTR-R 1 vs 0.95 (t $(1998) \geq 8.93$, $p < 0.001$).

These results are not surprising given that different carrier(family) types have high dissimilarity between them. Also, samples belonging to the same carrier have on average 60% of similar attributes. Figure 7 shows the similarity between the carrier types. The similarity between families is calculated in the same way as ACTR-IB partial matching with 0 indicating complete match while -1 complete mismatch.

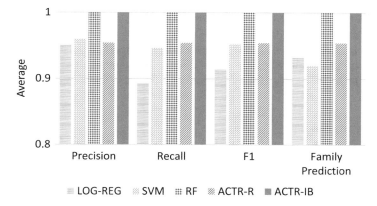

Fig. 6 Average precision, recall, F1 and family prediction comparisons for LOG-REG, SVM, RF, ACTR-R and ACTR-IB for different carrier samples

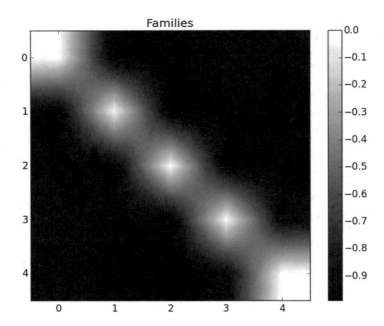

Fig. 7 Similarity matrix for five different carriers

Different Carriers-Mutation

For this case, we generate the same samples as in the previous experiment but with maximum mutation between samples belonging to the same carrier. We generated 1000 samples for each carrier with maximum mutation. In this case ACTR-IB had an average F1 of 1 outperforming LOG-REG 1 vs 0.83, SVM 1 vs 0.88, RF 1 vs 0.96 and ACTR-R 1 vs 0.92 (t (1998) $\geq$ 7, $p < 0.001$) (Fig. 8). Also for family prediction ACTR-IB outperformed LOG-REG 1 vs 0.85, SVM 1 vs 0.88, RF 1 vs 0.95 and ACTR-R 1 vs 0.92 (t (1998) $\geq$ 7, $p < 0.001$).

High mutation induces high variance between samples associated with the same carrier making the classification task difficult. High mutation samples belonging to same carrier have only 20% of common attributes as compared to 60% for low mutation.

Less Training Data

In order to see how the cognitive models perform with less training data, we repeated the different-carrier mutation experiment with 10% of the training data selected uniformly at random (300 samples). Even with less training data ACTR-IB had an average F1 of 0.93 outperforming LOG-REG 0.93 vs 0.71, SVM 0.93 vs 0.6, RF 0.93 vs 0.83 and ACTR-R 0.93 vs 0.88 (t (1998) $\geq$ 2.89, $p \leq 0.001$) (Fig. 9). Also for family prediction ACTR-IB outperformed LOG-REG 0.91 vs 0.73 (t (1998) = 19.3, $p < 0.001$), SVM 0.91 vs 0.58, RF 0.91 vs 0.79 and ACTR-R 0.91 vs 0.88 (t (1998) $\geq$ 2.05, $p \leq 0.04$).

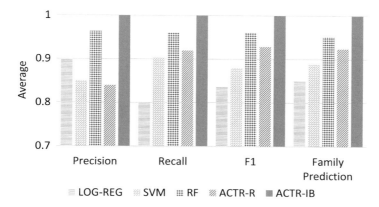

Fig. 8 Average precision, recall, F1 and family prediction comparisons for LOG-REG, SVM, RF, ACTR-R and ACTR-IB for different carrier mutated samples

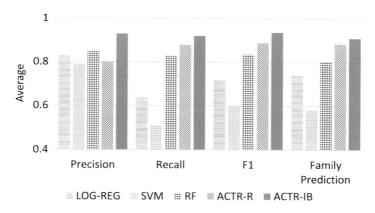

Fig. 9 Average precision, recall, F1 and family prediction comparisons for LOG-REG, SVM, RF, ACTR-R and ACTR-IB for less training data

Different Carriers: Low-High Mutation

For this case, we consider the low mutation samples as training data and the high mutation samples as testing. Figure 10 shows the comparison results. ACTR-IB had an average F1 of 0.96 outperforming LOG-REG 0.96 vs 0.83, SVM 0.96 vs 0.92, RF 0.96 vs 0.93 and ACTR-R 0.96 vs 0.88 (t (2498) ≥ 15.7, $p < 0.001$) (Fig. 10). Also for family prediction ACTR-IB outperformed LOG-REG 0.96 vs 0.81, SVM 0.96 vs 0.92, RF 0.96 vs 0.94 and ACTR-R 0.96 vs 0.88 (t (2498) ≥ 7, $p < 0.001$).

Leave One Carrier Out Cross-Validation

To see how the models generalize to unseen malware family(carrier), we performed a leave-one-carrier-out comparison, where we test the models against one previously unseen malware carrier. ACTR-IB performs better or on par with all other baseline

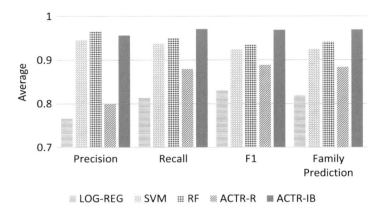

Fig. 10 Average precision, recall, F1 and family prediction comparisons for LOG-REG, SVM, RF, ACTR-R and ACTR-IB for low-high mutated samples

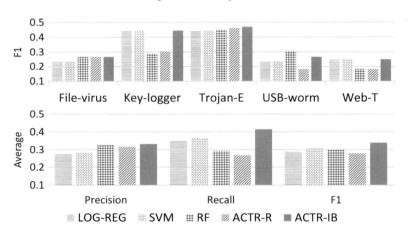

Fig. 11 Average F1 values for five malware carriers (above) and the average precision, recall and F1 across all carriers (below) for LOG-REG, SVM, RF, ACTR-R and ACTR-IB for leave-one-carrier-out

approaches for all the carriers. It clearly outperforms all the approaches in recalling most of the actual tasks (40%) (Fig. 11). ACTR-IB has shown to generalize for unseen malware families [10]. This case is difficult given the fact that the test family is not represented during training, hence task prediction depends on associating the test family with the training families that perform similar tasks.

Same Carrier

As seen in the previous experiments, different carrier types makes the task easier because of less similarity between them. We now test the performance, on same carrier type performing exactly one task. Since there are 17 tasks in the GVDG

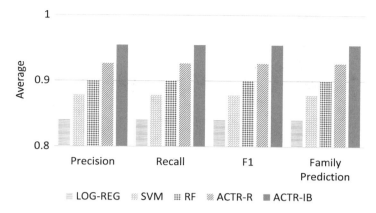

Fig. 12 Average precision, recall, F1 and family prediction comparisons for LOG-REG, SVM, RF, ACTR-R and ACTR-IB for unencrypted same carrier samples

tool, we generate 100 samples for each task for carrier type File-virus. In this experiment each task represents one family. Thus in total we have 1700 samples. We do the 60–40 split experiment. From Fig. 12 ACTR-IB had an average F1 of 0.95 outperforming LOG-REG 0.95 vs 0.84, SVM 0.95 vs 0.87, RF 0.95 vs 0.90 and ACTR-R 0.95 vs 0.92 (t (678) $\geq$ 1.52, $p \leq 0.13$). Since each family performs exactly one task the family prediction is similar to F1. Using the same carrier for each payload makes the task difficult as can be seen from the similarity matrix for the 17 payloads (Fig. 13).

Same Carrier-Encryption
The GVDG tool provides the option for encrypting the malware samples for the File-virus carrier type. We use this option to generate 100 encrypted malware samples for each task(payload) and use them as test data with the unencrypted versions from the same carrier experiment as training samples. From Fig. 14 ACTR-IB had an average F1 of 0.9 outperforming LOG-REG 0.9 vs 0.8, SVM 0.9 vs 0.8, RF 0.9 vs 0.74 and ACTR-R 0.9 vs 0.88 (t (1698) $\geq$ 2.36, $p \leq 0.02$). Encrypting malware samples morphs the task during execution making it difficult to detect during analysis. Hence the drop in performance as compared to non-encrypted samples. We note that SVM performs better than RF likely because it looks to maximize generalization.

Runtime Analysis
Table 5 shows the classifier run times for the experiments. Machine learning techniques are faster but have large training times, which increase almost linearly with the size of the knowledge base. Hence updating the knowledge base is computationally expensive for these methods, as it has to re-estimate the parameters every time. The same notion holds true for ACTR-R, since computing the rules during training phase is expensive as can be seen from the large training times. ACTR-IB on the other hand, has no explicit training phase, so the only time cost

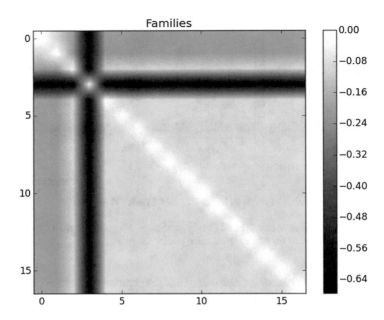

Fig. 13 Similarity matrix for 17 versions of the same carrier

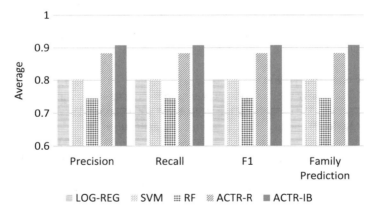

Fig. 14 Average precision, recall, F1 and family prediction comparisons for LOG-REG, SVM, RF, ACTR-R and ACTR-IB for encrypted same carrier samples

is during testing. In fact ACTR-IB is faster than SVM and RF for same/encrypted carrier experiments.

Scaling of Instance-Based Model

Finally to conclude the GVDG experiments, we run ACTR-IB on a combination of all the above variations of dataset to highlight the space requirements for the learning model. The dataset comprises of five different carriers with low/high muta-

Table 5 Classifier run times

Experiment	Model	Train (s)	Test (s)
Different carriers	LOG-REG	202	7
	SVM	250	50
	RF	280	30
	ACTR-R	6443	143
	ACTR-IB	–	453
Mutated carriers	LOG-REG	214	18
	SVM	260	63
	RF	303	85
	ACTR-R	7223	185
	ACTR-IB	–	465
Same carriers	LOG-REG	152	4.22
	SVM	270	38
	RF	290	55
	ACTR-R	4339	120
	ACTR-IB	–	205
Encrypted carriers	LOG-REG	180	15
	SVM	300	80
	RF	353	110
	ACTR-R	6103	180
	ACTR-IB	–	365

tion (10,000 samples) and same carrier encrypted/non-encrypted (3400 samples). Based on the tasks they perform we have in total 22 families represented by 13,400 samples. The analysis reports generated by cuckoo take up 4 GB of disk space for the samples. We significantly reduce the size to 600 MB by parsing the analysis reports and extracting attributes. We set aside 10% of the samples for testing (1340) and iteratively add 10% of the remaining data for training. Table 6 gives a summary of the average F1 measure and testing time for ACTR-IB. The results are averaged across ten trials. There is a steady increase in performance till we reach 40% of the training data, after that the F1 measure remains almost constant. This experiment clearly indicates the ability of the ACTR-IB to learn from small amount of representation from each family, significantly reducing the size of the knowledge base required for training. We are also looking into techniques to reduce the time requirements of instance-based learning algorithm (e.g., Andrew Moore explored efficient tree-based storage). There are also techniques for reducing space requirements, [27] merged training instances in the ACT-R-Gammon model and obtained considerable space savings at little performance cost.

Table 6 Summary of ACTR-IB results

Fraction of training data	F1 measure	Test time (s)
0.1	0.77	418
0.2	0.82	839
0.3	0.90	1252
0.4	0.97	1676
0.5	0.97	2100
0.6	0.97	2525
0.7	0.97	2956
0.8	0.98	3368
0.9	0.98	3787
1.0	0.98	4213

5.3   MetaSploit

MetaSploit is a popular penetration testing tool used by security professionals to identify flaws in the security systems by creating attack vectors to exploit those flaws [28]. Penetration testing may also be defined as the methods an attacker would employ to gain access to security systems. Hence identifying the tasks the exploit was designed to perform is important to counter the exploit.

For this experiment we generate exploits that attacks windows operating systems. Each exploit has a set of tasks associated with it. The tasks include setting up tcp & udp back-door connections, adding unauthorized users to the system, modifying root privileges, download executables and execute them on the local machine, prevent writing of data to disk, deleting system folders, copying sensitive information etc. We generated 4 exploit families with 100 samples each performing on average 4 tasks. We induced mutation between samples belonging to the same family making the classification task difficult. We perform a 60–40 split training-testing experiment and average the results across ten trials. From Fig. 15, ACTR-IB had an average F1 of 0.86 outperforming LOG-REG 0.86 vs 0.62, SVM 0.86 vs 0.82, RF 0.86 vs 0.82, ACTR-R 0.86 vs 0.81 and INVINCEA 0.86 vs 0.8 (t $(158) \geq 1.94$, $p \leq 0.05$). Also for family prediction ACTR-IB outperformed LOG-REG 0.8 vs 0.7, SVM 0.8 vs 0.72, RF 0.8 vs 0.72 and ACTR-R 0.8 vs 0.71 (t $(158) \geq 2.53$, $p \leq 0.01$).

5.4   Discussion

We evaluated the two proposed cognitive models on three different datasets under various operational conditions (mutated and encrypted malware samples). The instance based model performs on par or better than the rule-based model and standard machine learning approaches. The performance improvement can be attributed to different ACT-R modules (partial matching and spreading activation) that model different aspects of the malware sample. Partial matching computes the

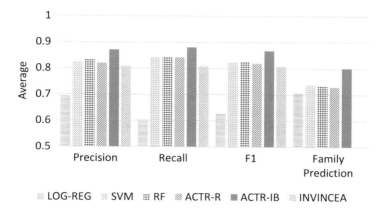

Fig. 15 Average precision, recall, F1 and family prediction comparisons for LOG-REG,SVM, RF, ACTR-R and ACTR-IB for metasploit samples

similarity between malware samples, while spreading activation identifies attributes that are indicative of a given malware family and the tasks that it is designed to perform. Hence, in cases where the training data is significantly less (as in one of the GVDG experiments), the proposed model is able to identify attributes representative of a particular malware family making the goal of correct task prediction better as compared to standard machine learning approaches which do require more training data for better generalization.

Experiments on the GVDG dataset under conditions of mutation and encryption provide further insights in the working of the cognitive models. For mutation, the malware samples used for training differ significantly from the ones used for testing. In this case the partial matching module does not contribute much towards the activation function but in turn the spreading activation is able to identify attributes that represent tasks that the malware sample is designed to perform thus making the correct task prediction in majority of the test cases. A similar behavior is observed for the encryption experiment as well. For the GVDG experiment with less training data shows how well the cognitive models are able to generalize with less training data which is difficult for standard machine learning approaches.

5.5 Task Prediction from Hacker Activities

In all the experiments discussed so far, the tasks associated with a given piece of malware are predefined and do not change with time. In this section, we try to map the tasks that a hacker is trying to achieve from the activities it performs on a compromised system. For the entire experiment only one malware is used whose sole purpose is to create a tcp backdoor connection to let the hacker have access to the system. We evaluate the test samples only using ACTR-IB and not other machine

Table 7 Summary of
ACTR-IB results

Subject	Average precision	Average recall	Average F1
Hacker-1	0.8	0.85	0.83
Hacker-2	0.85	0.85	0.85

learning methods. The goal of this experiment is to demonstrate how the system can deal with real time hacker activities on a compromised system.

The experimental setup is as follows. We keep the Cuckoo sandbox running on the system by executing the malware. This will create a connection between the hacker and the system. Once the hacker gains control of the machine, he can perform operations in order to achieve his objectives. We treat these objectives as the tasks that the hacker wants to complete on the system. Once these tasks are completed, Cuckoo generates an analysis report detailing the behavioral analysis of the hacker. However, these analysis are too detailed and do not provide a clear picture of the main tasks of the hacker on the machine. Hence, traditionally, this will often require an expert security analyst to go through large analysis results to determine the task, which is often time consuming. But instead we can feed the analysis report to the ACTR-IB model to get a prediction of the hacker tasks. For this experiment we use the Metasploit dataset discussed in Sect. 5.3 as the knowledge base for the instance based approach. For the test set we generate samples in real time with hackers trying to achieve their goals (tasks) on the compromised system. This test also illustrates how well our model generalizes, as we are identifying hacker behavior using historical data that was not generated by the hacker—or even a human in this case. We consider two hackers, who are given a list of the payloads (tasks) to complete from the list mentioned in Sect. 5.3. They always perform a fraction of the tasks assigned to them at a given time instance and then the model is tested on predicting these tasks.

We generate ten such attacks, five from each hacker. Each attack consists of achieving five tasks on average. We note that for each of the test sample the malware used is the same. ACTR-IB results are presented in Table 7. The results are averaged for each hacker across test samples. Table 8 shows the actual and predicted tasks for Hacker-1 for five different attack instances. The results for Hacker-2 were analogous.

6   Related Work

Identification of Malicious Software   The identification of whether or not binary is malicious [29, 30] is an important related, yet distinct problem from what we study in this chapter and can be regarded as a "first step" in the analysis of suspicious binaries in the aftermath of a cyber-attack. However, we note that as many pieces of malware are designed to perform multiple tasks, that successful identification of a

Table 8 Actual and predicted Hacker-1 attacks

Attack instance	Actual tasks	Predicted tasks
1	Setup backdoor connection modify root privileges uninstall program copy files	Setup backdoor connection modify root privileges uninstall program delete system files prevent access to drive
2	Setup backdoor connection modify root privileges download executables execute files copy files	Setup backdoor connection modify root privileges download executables execute files delete files
3	Setup backdoor connection modify root privileges add unauthorized users start keylogging uninstall program delete files prevent access to drives	Setup backdoor connection modify root privileges add unauthorized users start keylogging uninstall program delete files
4	Setup backdoor connection add unauthorized users prevent writing data to disk delete files copy files	Setup backdoor connection add unauthorized users prevent writing data to disk delete files modifying root privileges prevent access to drives
5	Setup backdoor connection download executables execute files start keylogging	Setup backdoor connection download executables execute files start keylogging

binary as malicious does not mean that the identification of its associated tasks will be a byproduct of the result.

Malware Family Classification There is a wealth of existing work on malware family identification [2–4, 6, 31–33]. The intuition here is that by identifying the family of a given piece of malware, an analyst can then more easily determine what it was designed to do based on previously studied samples from the same family. However, malware family classification has suffered from two primary drawbacks: (1) disagreement about malware family ground truth as different analysts (e.g. Symantec and McAfee) cluster malware into families differently; and (2) previous work has shown that some of these approaches mainly succeed in "easy to classify" samples [5, 6], where "easy to classify" is a family that is agreed upon by multiple malware firms. In this chapter, we infer the specific tasks a piece of malware was designed to carry out. While we do assign malware to a family as a component of our approach, it is not the focus of our comparison (though we show family prediction results as a side-result). Further, we also describe and evaluate a variant of our instance-based method that does not consider families and yields a comparable performance.

Malware Task Identification With regard to direct inference of malware tasks, the major related work includes the software created by the firm Invincea [7] for which we have included a performance comparison. Additionally, some of the ideas in this chapter were first introduced in [10–12]. However, these work primarily focused on describing the intuitions behind the cognitive modeling techniques and only included experimental evaluation on two datasets (the Mandiant APT1 and GVDG datasets). The experimental evaluation in this chapter includes additional experiments for the GVDG dataset to consolidate the previous experiments. Also algorithm analysis and parameter exploration are provided for the cognitive models. In addition we introduce a popular penetration tool used by security analyst Metasploit and present new results on this tool.

7 Conclusion

In this chapter, we introduced an automated method that combines dynamic malware analysis with cognitive modeling to identify malware tasks. This method obtains excellent precision and recall—often achieving an unbiased F1 score of over 0.9— in a wide variety of conditions over three different malware sample collections and two different sandbox environments—outperforming a variety of baseline methods.

Currently, our future work has three directions. First, we are looking to create a deployed version of our approach to aide cyber-security analysts in the field. Second, we look to enhance our malware analysis to also include network traffic resulting from the sample by extending the capabilities of the sandbox. Finally, we also look to address cases of highly-sophisticated malware that in addition to using encryption and packing to limit static analysis and employ methods to "shut down" when run in a sandbox environment [34]. We are exploring multiple methods to address this such as the recently introduced technique of "spatial analysis" [35] that involves direct analysis of a malware binary.

References

1. M. Sikorski, A. Honig, *Practical Malware Analysis: The Hands-On Guide to Dissecting Malicious Software*, 1st edn. (No Starch Press, San Francisco, 2012)
2. U. Bayer, P.M. Comparetti, C. Hlauschek, C. Kruegel, E. Kirda, Scalable, behavior-based malware clustering, in *NDSS*, vol. **9** (Citeseer, 2009), pp. 8-11
3. J. Kinable, O. Kostakis, Malware classification based on call graph clustering. J. Comput. Virol. **7**, 233–245 (2011)
4. D. Kong, G. Yan, Discriminant malware distance learning on structural information for automated malware classification, in *Proceedings of the 19th ACM SIGKDD. KDD '13, New York* (ACM, New York, 2013), pp. 1357–1365
5. P. Li, L. Liu, D. Gao, M.K. Reiter, On challenges in evaluating malware clustering, in *International Workshop on Recent Advances in Intrusion Detection* (Springer, Berlin, 2010), pp. 238–255
6. R. Perdisci, ManChon, Vamo: towards a fully automated malware clustering validity analysis, in *Proceedings of the 28th Annual Computer Security Applications Conference* (ACM, New York, 2012), pp. 329–338
7. Invencia, Crowdsource: crowd trained machine learning model for malware capability detection (2013). http://www.invincea.com/tag/cynomix/

8. Kaspersky, Gauss: abnormal distribution (2012). https://media.kasperskycontenthub.com/wp-content/uploads/sites/43/2018/03/20134940/kaspersky-lab-gauss.pdf
9. C. Lebiere, P. Pirolli, R. Thomson, J. Paik, M. Rutledge-Taylor, J. Staszewski, J.R. Anderson, A functional model of sensemaking in a neurocognitive architecture. Comput. Intell. Neurosci. 5:5–5:5 (2013)
10. C. Lebiere, S. Bennati, R. Thomson, P. Shakarian, E. Nunes, Functional cognitive models of malware identification, in *Proceedings of ICCM, ICCM 2015, Groningen, April 9–11* (2015)
11. R. Thomson, C. Lebiere, S. Bennati, P. Shakarian, E. Nunes, Malware identification using cognitively-inspired inference, in *Proceedings of BRIMS, BRIMS 2015, Washington DC, March 31–April 3* (2015)
12. E. Nunes, C. Buto, P. Shakarian, C. Lebiere, R. Thomson, S. Bennati, J. Holger, Malware task identification: a data driven approach, in *Proceedings of International Symposium on Foundation of Open Source Intelligence and Security Informatics (FOSINT-SI)* (IEEE, Piscataway, 2015)
13. Rapid7, Metasploit: penetration testing software (2003). http://www.metasploit.com/
14. D. McWhorter, APT1: exposing one of China's cyber espionage units (2013). http://Mandiant.com
15. GVDG, Generator malware GVDG (2011)
16. Mandiant, Mandiant APT1 samples categorized by malware families. Contagio Malware Dump (2013)
17. J.R. Anderson, D. Bothell, M.D. Byrne, S. Douglass, C. Lebiere, Y. Qin, An integrated theory of mind. Psychol. Rev. **111**, 1036–1060 (2004)
18. C. Gonzalez, J.F. Lerch, C. Lebiere, Instance-based learning in dynamic decision making. Cogn. Sci. **27**(4), 591–635 (2003)
19. R.S. Sutton, A.G. Barto, *Introduction to Reinforcement Learning*, 1st edn. (MIT Press, Cambridge, 1998)
20. T.J. Wong, E.T. Cokely, L.J. Schooler, An online database of ACT-R parameters: towards a transparent community-based approach to model development, in *Proceedings of the 10th International Conference on Cognitive Modeling* (Citeseer, 2010), pp. 282–286
21. D. Bothell, Act-r 6.0 reference manual (2004). http://act-r.psy.cmu.edu/actr6/reference-manual.pdf
22. L. Breiman, Random forests. Mach. Learn. **45**(1), 5–32 (2001)
23. C. Cortes, V. Vapnik, Support-vector networks. Mach. Learn. **20**, 273–297 (1995)
24. C.C. Chang, C.J. Lin, Libsvm: a library for support vector machines. ACM Trans. Intell. Syst. Technol. **2**(3), 27:1–27:27 (2011)
25. ISEC-Lab, Anubis: analyzing unknown binaries (2007). http://anubis.iseclab.org/
26. C. Guarnieri, A. Tanasi, J.B.M.S., Cuckoo sandbox (2012). http://www.cuckoosandbox.org/
27. S. Sanner, J.R. Anderson, C. Lebiere, M. Lovett, *Achieving Efficient and Cognitively Plausible Learning in Backgammon* (Carnegie Mellon University, Pittsburgh, 2000)
28. J. O'Gorman, D. Kearns, M. Aharoni, *Metasploit: The Penetration Tester's Guide* (No Starch Press, San Francisco, 2011)
29. I. Firdausi, C. Lim, A. Erwin, A.S. Nugroho, Analysis of machine learning techniques used in behavior-based malware detection, in *Proceedings of the 2010 Second International Conference on ACT. ACT '10, Washington, DC* (IEEE Computer Society, Philadelphia, 2010), pp. 201–203
30. A. Tamersoy, K. Roundy, D.H. Chau, Guilt by association: large scale malware detection by mining file-relation graphs, in *Proceedings of the 20th ACM SIGKDD. KDD '14* (ACM, New York, 2014), pp. 1524–1533
31. S.S. Hansen, T.M.T. Larsen, M. Stevanovic, J.M. Pedersen, An approach for detection and family classification of malware based on behavioral analysis, in *2016 International Conference on Computing, Networking and Communications (ICNC)* (IEEE, Piscataway, 2016), pp. 1–5
32. K. Sanders, X. Wang, Malware family identification using profile signatures. US Patent 9,165,142, 20 Oct 2015

33. C. Annachhatre, T.H. Austin, M. Stamp, Hidden Markov models for malware classification. J. Comput. Virol. Hacking Tech. **11**(2), 59–73 (2015)
34. M. Lindorfer, C. Kolbitsch, P. Milani Comparetti, Detecting environment-sensitive malware, in *Proceedings of the 14th International Conference on RAID. RAID'11* (Springer, Berlin, 2011), pp. 338–357
35. D. Giametta, A. Potter, There and back again: a critical analysis of spatial analysis (2014). https://archive.org/details/ShmooCon2014_A_Critical_Review_of_Spatial_Analysi

Social Media for Mental Health: Data, Methods, and Findings

Nur Shazwani Kamarudin, Ghazaleh Beigi, Lydia Manikonda, and Huan Liu

Abstract There is an increasing number of virtual communities and forums available on the web. With social media, people can freely communicate and share their thoughts, ask personal questions, and seek peer-support, especially those with conditions that are highly stigmatized, without revealing personal identity. We study the state-of-the-art research methodologies and findings on mental health challenges like depression, anxiety, suicidal thoughts, from the pervasive use of social media data. We also discuss how these novel thinking and approaches can help to raise awareness of mental health issues in an unprecedented way. Specifically, this chapter describes linguistic, visual, and emotional indicators expressed in user disclosures. The main goal of this chapter is to show how this new source of data can be tapped to improve medical practice, provide timely support, and influence government or policymakers. In the context of social media for mental health issues, this chapter categorizes social media data used, introduces different deployed machine learning, feature engineering, natural language processing, and surveys methods and outlines directions for future research.

Keywords Social media · Mental health · Online social network · Well-being

1 Introduction

Social media is a popular channel to spread information online. There are hundreds of millions of users that communicate with each other to share their thoughts, ideas, and personal experiences which overloads these channels with information. There are remarkable challenges when it comes to mental health issues [14]. A growing body of research have focused on understanding how social media activities can be use to analyze and improve the well-being of people, including mental health

N. S. Kamarudin (✉) · G. Beigi · L. Manikonda · H. Liu
School of Computing, Informatics and Decision Systems Engineering, Arizona State University, Tempe, AZ, USA
e-mail: nurkamarudin@asu.edu; gbeigi@asu.edu; lmanikon@asu.edu; huanliu@asu.edu

© Springer Nature Switzerland AG 2020
M. A. Tayebi et al. (eds.), *Open Source Intelligence and Cyber Crime*, Lecture Notes in Social Networks, https://doi.org/10.1007/978-3-030-41251-7_8

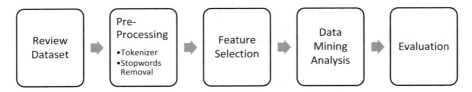

Fig. 1 Steps of social media analysis

[23, 34, 40]. With the presence of social media data, it is now easier to study the trend of mental health problems and also to help researchers get information from social media to study mental health issues [14]. The easy access and use of social media allow users to update their social media profiles without time or space restriction [45]. This makes social media a preferable medium for researchers for their investigations. In addition, it is cost-effective for information seekers. Users can easily get health-related information [32].

1.1 Social Media and Its Analysis

Social media has become a good source for data collection. There are different types of data that can be used from social media such as text, image, video, and audio. The amount of data on social media data increases rapidly. For example, on Twitter, 350,000 tweets are generated per minute and 500 million tweets are generated per day. A major factor that might affect social media users is the way they use social media because it can be very beneficial or toxic at the same time. For instance, active use of social media with two-way communication can be very beneficial to the user but it can also be destructive or toxic to the user [57].

Figure 1 shows the common steps in social media analysis. It starts with dataset review in which the researchers need to choose the right dataset for their experiment. The second step is data pre-processing, which means preparing the data for the experiment such as removing stop words or word/sentence tokenizing. The next step is to select meaningful features from social media data such as an image or textual features. After selecting the right features, it is data mining analysis which includes deploying various techniques to develop the desired model. The final step is an evaluation, employing different metrics such as accuracy, recall, precision, F1 scores, for example.

1.2 Mental Health Problem

Mental health has become a public concern nowadays. People have started to think about the importance of mental health problems and their effects on our society. This is not a minor issue; on the contrary, it is a very serious issue that can contribute to mental well-being. For example, graduate students nowadays may face anxiety, depression, and stress because of the high competition in the academic world, long work hours, and lack support from their advisor [12, 29]. These studies show that not just students [12, 25, 28, 29, 38], even employees [17, 18, 59] are all facing their own problems and authorities need to step up in order to help these groups of people. Researchers in the psychology field have studied this topic for decades. With the increasing use of social media data, this specific problem also attracts the attention of many computer scientists. Users actively share and communicate with online communities and researchers have found that it is a smart idea to leverage social media data to study this problem in order to help online communities and authorities at the same time. This can contribute to an immense change in overcoming this issue. Researchers have begun to investigate mental health problems such as post-traumatic stress disorder (PTSD) [15, 33, 46], depressive disorder [22, 43, 45, 52], suicidal ideation [16, 21, 37], schizophrenia [41], and anxiety [20, 27, 56] through social media data.

This chapter is organized into five sections. Section 2 discusses the social media data on mental health by categorizing the data into three categories, which are linguistic-, visual- and combined-based. Section 3 presents a set of approaches used by the researcher on mental health research. In addition, evaluation metrics and findings are discussed in Sect. 4. Lastly, the study is concluded and a list of possible future works is presented in Sect. 5.

2 Social Media Data on Mental Health

Various types of social media data have been used by researchers to study mental health. Existing work could be categorized in to three groups based on the type of utilized social media data, (1) linguistic-based data, (2) visual-based data, and (3) combination of linguistic and visual data. Next, we will discuss each separately.

2.1 Linguistic-Based Data

Over the past few years, research in crisis informatics have utilized language as a medium to understand how major crisis events unfold in affected populations, and how they are covered on traditional media as well as online media such as blogs and social media sites [55]. Studies have shown that social media can provide a comforting environment for support seekers especially when it comes to stigmatized issues that make them reluctant to share with individuals around them.

Social media has been accordingly used to understand users' mental health issues. Interesting work by Coppersmith et al. [16] studies suicide attempts or even ideation among users using their posts in Twitter. The authors crawl tweets from 30 geolocations from all over the United States with at least 100 tweets per location.

Then, they use natural language processing techniques to compare behavior of users who attempted suicide with those users who have previously stated that they were diagnosed with depression and neurotypical controls. In another work, Park et al. [45] focus on studying the impact of online social networks (OSN) on the depression issue. Accordingly, they collect two different kinds of data: (1) data from Internet-based screening test, which includes information of 69 participants who were asked to complete a questionnaire including depression related questions, and (2) collected tweets from Twitter from June 2009 to July 2009 that include the keyword 'depression'. After qualitative and quantitative analysis of collected data, authors show that social media data could be used to understand users' mental health issues.

Nadeem [43] use social media data to study Major Depressive Disorder (MDD) issue in individuals. They use a publicly available dataset that was built from Shared Task organizers of Computational Linguistics and Clinical Psychology (CLPsych 2015) [15]. This dataset includes information of Twitter users who were diagnosed with depression. In another work, Amir et al. use Twitter to study depression and post-traumatic stress disorder (PTSD) [2]. They investigate the correlation between users' posts and their mental state. In particular, they investigate if tweets could be used to predict whether a user is affected by depression and PTSD or not. De Choudhury et al. [22] also show that social media can be utilized for predicting another mental health condition, i.e., major depressive disorder (MDD), using Twitter data. In this work, authors employ crowd-sourcing technique to provide the ground truth for their experiments. In another work [27], authors use Twitter data to study gender-based violence (GBV). The authors use Twitter's streaming API to sample a set of GBV related tweets based on a set of key phrases defined by the United Nations Population Fund (UNFPA) [50].

Another well-known social media platform is Reddit, which is a forum-based social media platform that captures communication between the original post and the user who left a comment on the thread. Each thread discusses a specific topic, which is known as a "subreddit". The work by De Choudhury and De [20] uses Reddit to investigate how users seek mental health related information on online forums. They crawl mental health subreddits using Reddit's official API[1] and Python wrapper PRAW.[2] In another work, De Choudhury and Kıcıman [21] use Reddit to study the language style of comments left by users on the discussion forum in terms of influences towards suicidal ideation. This work fills the gap of how online social support can contribute to this specific problem. Authors use stratified propensity score analysis to determine if the user was affected by comments or not. They also estimate the likelihood that a user will receive a treatment based on the user's covariates. There was a study by De Choudhury et al. [24] with work specifically on mental health subreddits such as r/depression, r/mentalhealth, r/bipolarreddit, r/ptsd, r/psychoticreddit. Based on the time stamp, data were divided

---

[1]http://www.reddit.com/dev/api.

[2]https://praw.readthedocs.org/en/latest/index.html.

into a treatment group and control group. These groups were further divided based on the causal analysis in order to analyze the effect of comments on the content of earlier posts with comment shared and received by users in their dataset.

Another work by Saha and De Choudhury [55] uses Reddit data to study the effect of gun violence on college students and the way they express their experience on social media. Authors collect related data from Reddit. Then they develop an inductive transfer learning approach [44] to see the pattern of stress expression by computing the mean accuracy value. In particular, they first build a classifier which labels the expressed stress in posts as *High Stress* and *Low Stress*. Then, they adopt the trained classifier to categorize collected posts from Reddit in order to identify posts that express higher stress level after the shooting incident. Another work from Lin et al. [39] focuses on how online communities can affect the development of interaction within social media. This work investigates if new members will bring interruption in terms of perspectives towards social dynamic and lower content quality. The authors accordingly generate three questions related to user reception, discussion content and interaction patterns. They use Reddit data from Google BigQuery by choosing the top ten subreddits between years 2013 and 2014. They study user reception, post content, and commenting patterns among Reddit users. This work studies the role of online social support based on historical data in conjunction with its effect on future health. It also investigates linguistic changes in online communities over time using data from two peer reviewing communities. Cohan et al. [13] study mental health by analyzing the content of forum posts based on the sign of self-harm thoughts. Their main goal was to study the impact of online forums on self-harm ideation. They consider four level of severity for the post content. Author build a model that includes lexical, psycholinguistic, contextual and topic modeling features. Their data were collected from a well-known mental health forum in Australia, namely, ReachOut.com.[3]

2.2 Visual-Based Data

The popularity of visual-based social media has increased rapidly. Users tend to communicate on social media by posting their photographs. Photo sharing provides a unique lens for understanding how people curate and express different dimensions of their personalities [3]. People use photos to define and record their identity, maintain relationships, curate and cultivate self-representation, and express themselves [62].

Posting pictures have become one way of communication among social media users. The definition of "selfie" is a picture that users take of themselves. Kim et al. [36] study the behavior of selfie-posting using Instagram data. This work uses selfie-posting to predict the intention of users who post selfies on social networking

---

[3]https://au.reachout.com/.

sites (SNSs). This work defines five hypotheses before they start designing the experiment. Those hypotheses were: attitude toward the behavior of selfie-posting, subjective norm, perceived behavioral control, and narcissism, which were possibly related to the intention to post selfies on social network sites (SNSs). The last hypothesis was the intention to post selfies on SNSs is positively related to the actual selfie-posting behavior on SNSs. They begin with 89 Instagram users. They recruit these users based on their agreement to be part of the study. Two coders analyze each user's account and the total sample size was ($n = 85$). From the total number of participants, 9 were males and 76 were females. They also count the total number of the pictures posted on each user's account in a 6-week timestamp. Each user was required to answer a list of questions that were related to the standard Theory of Planned Behavior (TPB) variables such i.e., attitude, subjective norm, perceived behavioral control, and future intention based on [1].

Another work by Reece et al. [52] uses visual data for studying mental health in social media. The authors use Instagram data from 166 individuals with 43,950 photographs. In order to study the markers of depression, they use machine learning tools to categorize users to healthy and depressed groups. They began their experiment by crawling all posts from each user's account upon their agreement. Participant users were also required to answer a depression related questionnaire that contained specific questions based on inclusion criteria. In the last step of experiments crawled Instagram photos are rated using a crowd-sourced service offered by Amazon's Mechanical Turk (AMT) workers.

2.3   Combined Data of Linguistic and Visual-Based

Apart from using only visual data or only linguistic data, researchers also combine these two kinds of data to study social media influence on mental health. Sociologists also claim that it is not possible to communicate by using the only words; people also use pictures to communicate with each other [8]. A work by Burke et al. [10] used different features in Facebook data such as wall posts, comments, "likes", and consumption of friends' content, including status updates, photos, and friends' conversations with other friends in order to study the role of directed interaction between pairs. This work distinguish between two types of activity: directed communication and consumption. To do this, they recruited 1199 English-speaking adults from Facebook to be their research participants.

Andalibi et al. [3] study depression-related images from Instagram. They use image data and the matching captions to analyze if a user was having a depression problem or had faced this kind of problem in the past. They were keen on investigating if this group of users engage in a support network or not, and how social computing could be used to encourage this kind of support interaction among users. They gather 95,046 depression tagged photos posted by 24,920 unique users over 1 month (July 2014) using Instagram's API. All public details of each image were stored from these photos, such as user ID, number of likes and comments,

Table 1 List of mental health related social media data and available datasets

Type of data	Paper	Name	Availability
Linguistic data	[16, 43]	Computational Linguistic and Clinical Psychology (CLPsych)	Publicly available at https://bit.ly/2T0hMGO
	[39, 55]	Google News dataset	Publicly available at https://bit.ly/1LHe5gU
	[20, 27, 43, 45]	Reddit, Twitter, Facebook	Crawled using provided API by social media platform
Image data	[36]	Instagram	Publicly available at https://bit.ly/2SV5cbY
Combination of text and image data	[3, 40]	Twitter, Facebook, Instagram	Crawled using provided API by social media platform
Link data	[22, 35]	Twitter, Facebook	Crawled using provided API by social media platform

date/time of creation, and tags. After conducting data collection, they begin the experiment by analyzing images and their textual captions. They also develop a codebook that includes 100 sample images and captions. Those coders then manually discuss the codebook in order to provide the best result for the experiment. Then, they add 100 more sample images and repeat the same steps.

Likewise, Peng et al. [48] investigate the effect of pets, relationship status, and having children, towards user happiness using Instagram pictures and captions. They use several hashtags such as #mydog, #mypuppy, #mydoggie, and #mycat, #mykitten, #mykitty to gather images of pet owner from Instagram. For non-pet owners, hashtags such as #selfies, #me, and #life were used to crawl the data. Before they started with the experimental steps, they began by classifying their data. The authors also provided the processed human face data called the face library for other researchers to use (see Table 1).

Manikonda and De Choudhury [40] use popular image-based media data in order to study mental health disclosure. This work extracts three main visual features from each image that they have in the corpus. Those features include visual features (e.g., color), themes, and emotions. Authors crawl the data from Instagram using Instagram's official API.[4] The main focus of this work was to study mental health disclosure based on visual features, emotional expression and how visual themes contrast with the language in a social media post. They specifically choose ten mental health challenges from Instagram before they crawl two million public images and textual data from that particular medium. Those categories contain ten types of disorders, which were: anxiety disorder, bipolar disorder, eating

---

[4]https://www.instagram.com/developer/.

disorder, non-suicidal self-injury, depressive disorder, panic disorder, OCD, PTSD, suicide, and schizophrenia. Before they begin their experiment, they consult with the Diagnostic and Statistical Manual of Mental Health Disorders (DSM-V) in order to confirm that their final disorder categories could be reliable.

To sum-up, user-generated social media data is heterogeneous and consists of different aspects such as text, image, and link data. Table 1 summarizes different datasets used by researchers to study mental health using social media.

3 Studying Mental Health in Social Media

As social media data has recently emerged as the main medium to spread information among online communities, there are also various approaches used by researchers to study related problems. This section discusses the approaches for studying mental health on social media. In this section, we elaborate on types of techniques or tools used in their research. Figure 2 shows the categorization of social media analysis, namely machine learning methods, feature engineering and survey methods. Next, we introduce how social media analysis are used for mental health analysis in social media.

3.1 Machine Learning Methods

We discuss machine learning methods in terms of classification, clustering, and prediction with social media data for studying mental health problems.

Classification

In order to estimate the likelihood of having depression among users within a dataset, a work by Nadeem [43] employ four types of classifiers (Decision Trees, Linear Support Vector Classifier, Logistic Regression, and Naive Bayes). They present a set of attributes to characterize the behavioral and linguistic differences of two classes. To do that, author utilize scikit-learn[5] which is a popular tool with many supervised and unsupervised machine learning algorithms [47].

---

[5]http://scikit-learn.org/stable/.

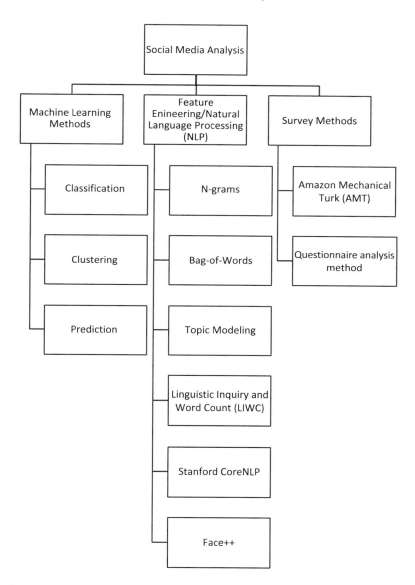

Fig. 2 Social media analysis approaches

Clustering

De Choudhury et al. [22] cluster the ego-networks among users on social media. By clustering the ego-network, they study the characteristics of the graphs based on the egocentric measures, such as the number of followers, number of followees, reciprocity, prestige ratio, graph density, clustering coefficient, 2-hop neighborhood, embeddedness and number of ego components. Another work by [35] studies the correlation of social ties and mental health and finds that depressed individuals tend to cluster together.

Prediction

The work by Reece et al. [52] predicts depression using photographic details, such as color analysis, metadata components and algorithmic face detection. In another approach, the authors in [22] divide users into two categories based on differences in behavior. For each user, they utilize a set of behavioral measures, such as mean frequency, variance, mean momentum, and entropy based on a user's 1-year Twitter history. In order to avoid overfitting, the authors employ principal component analysis (PCA), then compared their method with several different parametric and non-parametric classifiers.

3.2 Feature Engineering/Natural Language Processing (NLP) Methods

Natural language processing plays a very important role in linguistic social media analysis. This subsection discusses feature representation techniques used in studying social media for mental health.

N-Grams

This text representational technique is widely adopted and is basically a set of co-occurring words within a given window. Features extracted using this technique are based on word frequency counts. In [20], authors calculate the most frequent unigrams from all Reddit posts and use negative binomial regression as their prediction model.

Saha et al. built a supervised machine learning model in order to classify stress expression in social media posts into binary labels of *High Stress* and *Low Stress* [55]. To develop the transfer learning framework for their experiment, they measure the linguistic equivalence by borrowing a technique from domain adaptation literature [19]. By using top n-grams ($n = 3$) as an additional feature, Saha et al. developed a binary Support Vector Machine (SVM) classifier to detect *High Stress* and *Low Stress*. To build their training set, the authors extract 500 *n*-grams from the Reddit posts that they crawled. They compute the cosine similarity and compare their data with a Google News dataset in a 300-dimensional vector space. Authors find that it is possible to use social media content to detect psychological stress. On the other hand, as Twitter data has a limited number of characters per post, another work [9] designs a character *n*-gram language model (CLM) to get the score for each short text. This specific method examines sequences of characters including spaces, punctuation, and emoticons. For example, if we have a set of data from two classes, the model is trained by recognizing the sequence of characters. Similar character sequences will be classified into the same class. Given a novel text, the model can do estimations on which class can produce and generate all the texts.

Furthermore, the authors in [40] extract *n*-grams ($n = 3$) to check the suitability and reliability of their corpus. Extracted *n*-grams have been further used to investigate if they are facing mental health disclosures. In order to extract visual features from the dataset, the authors pair OpenCV and Speeded Up Robust Features (SURF [6]). This approach is able to identify the meaningful themes from images. To study the linguistic emotions based on the visual themes, this work uses psycholinguistic lexicon LIWC and TwitterLDA. These two approaches help authors to measure the estimation of how themes and images were coherent to each other.

Bag-of-Words (BOW)

Bag-of-Words (BOW) is a basic text representation for texts widely used by researchers in this area. When implementing this approach, a histogram is created to indicate how often a certain word is present in the text [63]. A previous work by [60] shows that the bag-of-words approach can be useful to identify depression. The author in [43] utilizes word occurrence frequencies to quantify the content from Twitter data by assembling all words and measuring the frequency of each word. Similarly, [13] uses BOW to extract features from their dataset. This work also uses content severity of the users in order to help forum moderators to identify the critical users that are keen on committing self-harm.

Topic Modeling

One of the most mentioned topic modeling methods is Latent Dirichlet Allocation (LDA), which works by drawing distribution topics for each word in the document [7]. Then, words are grouped based on the distribution value. Similar words are in the same topic category. Cohan et al. [13] use the LDA model to find a set of topics from their data collection. By training the LDA topic model on the entire forum posts from their dataset, they are able to use the topic model as additional features for their experiment, which boosted the performance of their system and prove the effectiveness of topic modeling. Additionally, Manikonda and De Choudhury [40] use TwitterLDA to extract the linguistic themes from their dataset to see if visual and text are coherent to each other when it comes to mental health disclosure on Instagram.

Amir et al. [2] adopt a model known as Non-Linear Subspace Embedding (NLSE) approach [4] that can quantify user embedding based on Twitter post histories. The authors evaluate user embedding by using User2Vec (u2v), Para-graph2vec's PV-dm and PV-dbow models. They also leverage Skip-Gram in order to build vectors. Another design based on bag-of-embedding was bag-of-topics by using LDA to indicate topics presented in the user's posts [20]. They leverage LDA to identify types of social support on Reddit. They also consider information on practices that people share with the communities by characterizing self-disclosure in mental illness. The authors find that Reddit users discuss diverse topics. These discussions can be as simple as talking about daily routines but it can also turn into a serious discussion that involves queries on diagnosis and treatments.

Additionally, a work by Lin et al. [39] studies linguistic changes and for their data, the authors use several post-level measures including cross-entropy of posts and Jaccard self-similarity between adjacent posts. Then, the authors use the LDA model in order to compare the topic distribution among posts and general Reddit post samples. They also track the linguistic changes in sub-communities. In order to examine the interaction network's structural change, they calculate the exponent α in the network's power-law degree distribution which gives the graph densification of the network [11]. Reddit allow users to vote on each post and comments, and the authors leverage this feature by computing the average score and complaint comment percentage in order to investigate community reaction to the content produced by newcomers.

Linguistic Inquiry and Word Count (LIWC)[6]

It is a text analysis application that can be used to extract emotional attributes on mental health. This tool will be able to extract psycholinguistic features [13]. Manikonda and De Choudhury [40] use LIWC on texts associated with mental health images spanning different visual themes. LIWC can also characterize linguistic styles in posts from users [53]. Park et al. [45], use LIWC to quantify the level of depressive moods from their Twitter data. They compare a normal group vs. depressed group by measuring the average sentiment score from categories provided by the tools. LIWC contains a dictionary of several thousand words and each word fed to this tool will be scaled across six predefined categories: *social*, *affective*, *cognitive*, *perceptual*, *biological processes*, and *relativity*. Every criterion

---

[6]http://www.liwc.net/.

will have its own categories and sub-categories. For each sub-category, LIWC will assign specific scores for each word. The authors in [22] use LIWC to study Twitter users' emotional states. Then, they use point wise mutual information (PMI) and log-likelihood ratio (LLR) to extract more features from their corpus. ElSherief et al. [27] leverage LIWC in order to measure interpersonal awareness among users by differentiating perceived user and actual user characteristics.

A study by De Choudhury and De [20] captures the linguistic attributes of their data by measuring the unigram and then employ psycholinguistic lexicon LIWC. They choose LIWC because it can categorize Redditors' emotions. They also examine the factors that drive social support on mental health Reddit communities, where the authors build a statistical model by measuring the top most frequent semantic categories from LIWC. The authors in [21] adopt the LIWC lexicon to study the various sociolinguistic features from their dataset and then measure the t-tests in order to analyze the differences between subpopulations. Coppersmith et al. [16] also use LIWC to study the pattern of language in conjunction with psychological categories generated from their dataset. This work uses LIWC to interpret how language from a given psychological category will be scored by the classifiers that they built. Likewise, Saha and De Choudhury [55] investigate on quantifying the psycholinguistic characterization. They employ LIWC measures to understand the psychological attributes in social media.

Stanford CoreNLP

A study by Peng et al. [48] classifies the images and textual sentiments to see the significant role of those factors in reducing stress and loneliness among individuals. In order to interpret a user's happiness from a caption, they also utilize a sentiment analysis method, which is Valence Aware Dictionary and Sentiment Reasoner (VADER). Saha and De Choudhury [55] use Stanford CoreNLP's sentiment analysis model to retrieve the sentiment class of posts.

Face++

A deep learning-based image analysis tool that is very useful for the facial recognition research field. It is an open-source face engine built with the convolutional neural network (CNN). In the work by ElSherief et al. [27], the authors use Twitter user's profile picture to predict the demographic information of the user by using Face++ API. Based on the GBV content, they investigate user involvement in GBV related post by leveraging the language nuances of those posts. Face++ is also used in [48] to do an experiment on face analysis by extracting user's information such as demographics inference, user relationship status, if a user has children, and then analyze user happiness.

3.3 Survey Methods

This subsection discusses works that use human intelligence tasks (HITs) for their analysis process, namely Amazon's Mechanical Turk and questionnaire based analysis. We also discuss the tools used.

Amazon Mechanical Turk (AMT)

It is one of the widely used crowdsourcing platforms. On AMT, chunks of work are referred to as Human Intelligence Tasks (HIT) or micro-tasks [58]. This technique is leveraged in [20] to label words related to mental illness for their dataset. Moreover, Amazon's Mechanical Turk is used in [22] to take a standard clinical depression survey followed by several questions on depression history and demographics. The crowdworkers have the option to either include their *public* Twitter profiles or not in the analyzing process. Reece et al. use AMT service to rate the Instagram photographs collected for their experiment [52]. Raters are asked to judge how interesting, likable, happy, and sad each photo seemed, on a 0–5 scale.

Questionnaire Analysis Method

The questionnaire can be very helpful for researchers to get more insight on the topic they are studying. For example, authors in [36] use this technique to measure attitude towards selfie-posting on semantically differential scales (e.g., bad/good, pleasant/unpleasant). By using the Narcissism Personality Inventory (NPI) [30], they measure Narcissism and participants' respond on a seven-point Likert scale (1 = "strongly disagree" to 7 = "strongly agree"). The authors uses AMOS 22 to test their hypotheses and see the relationships between attitude, subjective norm, perceived behavioral control, and narcissism toward their main question, which was the user's intention and behavior toward posting selfies on SNSs. In another work [52], authors uses the Center for Epidemiologic Studies Depression Scale (CES-D) [51] to screen participants' depression level for the depressed user group. Qualified participants are asked to share their Instagram usernames and history. An app embedded in the survey allow participants to securely log into their Instagram accounts and share their data.

Burke et al. [10] run a survey to analyze the relationship of social well-being and SNSs activity. Each user is required to answer survey questions using the format from the UCLA loneliness scale [54], Likert scales, and Facebook intensity scales [26]. They analyze their data without analyzing the private data of users, such as friend networks or identifiable information. They measure the number of friends and time spent on SNSs for each user so that they can get the answer to their research questions. In another work[45], authors use the CES-D to measure depressive symptoms and [22] also use CES-D to determine the depression levels

Table 2 The summary of methods and tools in the application of mental health studies

Approach	Paper
Linguistic Inquiry and Word Count (LIWC)	[20, 22, 27, 40, 45, 55]
Latent Dirichlet Allocation (LDA)	[2, 7, 13, 20, 40]
N-grams	[9, 16, 20, 40, 55]
Amazon Mechanical Turk (AMT)	[20, 22, 52]
Center for Epidemiologic Studies Depression Scale (CES-D)	[22, 45, 52]

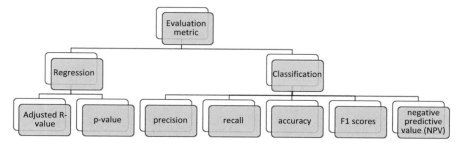

Fig. 3 Categories of evaluation metrics used for mental health analysis in social media

of crowdworkers by distributing a depression survey to Amazon Mechanical Turk. As a summary, Table 2 shows the methods and tools that we discuss in this section.

4   Evaluation Methods and Findings

In this section we discuss the utilized evaluation metrics and findings from the aforementioned papers. Figure 3 represents the utilized evaluation metrics for a mental health studies in social media. We begin this section by overviewing evaluation metrics. Then, we discuss the findings from previous works.

4.1   Evaluation Metrics

There are various evaluation metrics in data mining analysis. We discuss the evaluation metrics used for studying mental health using social media data. The most common prediction metrics include precision, recall, F1 scores, and recipient operating classification (ROC) curves. Equation 1 shows the calculation of precision defined as the number of true positives (TP) divided by the sum of all positive predictions, TP and false positive (FP). Equation 2 is defined as the number of TP divided by the sum of all positives in the set, TP and false negative (FN). Equation 3 measures F1 scores by considering both precision and recall. The F1

score is the harmonic average of the precision and recall, and an F1 score reaches
its best value at 1. Another metric known as adjusted R-squared is represented in
Eq. 4. Adjusted R-squared is often used for explanatory purposes and explains how
well the selected independent variable(s) explain the variability in the dependent
variable(s). In adjusted R-squared, n is the total number of observations and k is the
number of predictors. Adjusted R-squared is always less than or equal to R-squared.

$$precision = \frac{TP}{TP + FP} \tag{1}$$

$$recall = \frac{TP}{TP + FN} \tag{2}$$

$$F_1 = 2 \cdot \frac{precision \cdot recall}{precision + recall} \tag{3}$$

$$(R^2_{adj}) = 1 - [\frac{(1 - R^2)(n - 1)}{n - k - 1}] \tag{4}$$

Several works implement these metrics for their experiments [13, 22, 43, 55]. De
Choudhury et al. [22] evaluate their proposed classification approach by predicting
individual depression level from their posts. They use precision, recall, accuracy, and receiver-operator characteristic (ROC) for evaluation. Their experimental
results show that their classifier has a good performance in depression prediction.
Another work from De Choudhury and Kıcıman [21] measures the most positive
or negative z scores in order to differentiate between mental health with influence
risks to suicidal ideation (SW) users and mental health users. On the contrary, a
study by [45] uses the coefficients from regression models to predict the Center
for Epidemiologic Studies Depression Scale (CES-D) score. It then evaluates the
proposed approach by measuring adjusted R-squared (Eq. 4) and p-value. Another
work [2] measures *F1* and *binary F1* with respect to a mental condition in order to
measure the performance of different models for its experiment.

Furthermore, authors in [27] measure the favorite rate and retweet rate for each
tweet in order to count how many times a tweet was favorited and retweeted,
respectively. These metrics are used to explore the engagement of users with gender-
based violence (GBV) content on Twitter. Saha and De Choudhury [55] measure
accuracy, precision, recall, F1-scores and ROC-AUC in order to see the performance
level of their stress predictor classifier. Likewise, Coppersmith et al. [16] plot ROC
curve of the performance for distinguishing people who attempted suicide from their
age- and gender-matched controls. In order to compare the accuracy of all data and
pre-diagnosis in their model prediction, [52] measure recall, specificity, precision,
negative predictive value (NPV) and F1-scores.

Furthermore, a study by Manikonda and De Choudhury [40] calculates the
Spearman rank correlation coefficients to compare the most frequent tags across
all pairs of visual themes that belonged to six visual themes of mental health-related

posts. Burke et al. present the ordinary least square (OLS) regressions for bridging and bonding social capital and loneliness based on the overall SNS activities [10]. Additionally, [39] study the effect of newcomers to existing online forums such as Reddit. They leverage the regression analysis by calculating the adjusted R-squared in order to measure the average score of post voting and complaint comment percentage on the content of the subreddit. In another work, Andalibi et al. calculate Cohen's Kappa coefficient to analyze depression related images along with the textual captions [3].

4.2 Output and Findings

In this section, we discuss the outputs and findings from previous works based on the type of data used. As we discussed in Sect. 2, we categorize works based on data that the authors used in their experiments. Here, we first discuss the findings of existing works that used linguistic data. Then, we review findings from visual data. Finally, we discuss findings from using combined data.

Linguistic-Based Findings

By focusing on linguistic-based experiments, [45] concludes that people disclose not only depressed feelings, but also very private and detailed information about themselves such as treatment history. For participants suffering from depression, their tweets were found to have high usage of words related to negative emotions and anger. In the end, users with depression tend to post more tweets about themselves than typical users. Likewise, the work by De Choudhury et al. [22] demonstrates that Twitter can be used as a platform to measure major depression in individuals. To develop the prediction framework, the authors calculate four statistical values from the corpus including mean, variance, momentum, and entropy of selected features. Then, they compare these values between depression and non-depression classes. They find that individuals with low social activity tend to have a greater negative emotion, high self-attentional focus, increased relational and medicinal concerns, and heightened expression of religious thoughts. We can conclude that social activity does play an important role in individual mental well being. This group of people with low social activity has close-knit networks, which are normally highly embedded with their audience. The authors conclude that useful signals from social media can be used to characterize the onset of depression in persona by measuring their social activity and expression through their social networks. This kind of experiment shows how much social media can contribute to the body of knowledge in finding the solution and helping individuals in need.

Similarly, a work by De Choudhury and Kıcıman [21] reports that comments play an important role in terms of giving support, especially among mental health communities. They observe that users who receive support from the online forums are more socially active and engage with the communities. Moreover, Nadeem [43] show that Twitter can be used as a tool to predict MDD among users. The novelty of this research is the proposed text classification system which can classify if the tweets from users are depressive in nature or not. They conclude that social media can capture the individual's present state of mind. The text classification system is also effective because Twitter users are using this medium as a place to express their feelings. In addition, this study shows that it is reliable to use social media data for studying mental health related issues. Another study by Amir et al. [2] propose a novel model to extract users characteristics from their tweets known as user embeddings and further investigate their mental health status with respect to depression and PTSD. Their results show the correlation between captured embeddings and users' mental health condition.

In addition, ElSherief et al. [27] show that people discuss GBV-related issues on social medias. GBV-related hashtags help Twitter users to express their feelings, especially to share experiences and seek support. It has been shown that the most expressed emotion was anger. Another work demonstrates that communication between users plays an important role in mental well-being [13]. Their results show that when users are more active in communicating with other users, it helps to decrease the content severity. In another work, [39] compares the effects of growing online communities to the current network and showed that users perception remains positive and growth has an impact on users' attention. Authors also find that high levels of moderation helps to maintain the positive perception of community content after getting defaulted and that the communities' language do not become more generic or more similar to the rest of Reddit after the massive growth.

Saha and De Choudhury [55] show that posts published after gun-violence incident on college campuses include higher level of stress in comparison to those posted before incident. The authors also find that there is an increase in self-attention and social orientation when the campus population reduced. Also, more students were observed engaging in death-related conversations. In another work [16], researchers find that people who attempt suicide engage less in conversation, showing that this group of users had a smaller proportion of their tweets directed to other users. This work demonstrates how social media data can be beneficial for understanding mental health-related studies. De Choudhury et al. [20] also find that there is a variation in each type of social support. They also observe that posts related to self-attention, relationship and health issues received more attention from Reddit communities. Moreover, they find that negative posts get more attention compared to positive posts. By studying user feedback from all posts, they conclude that certain types of disclosures receives more social support from the online communities.

Visual-Based Findings

Kim et al. [36] use the theory of planned behavior (TPB) [1] for studying behavioral intention using social media data. They find that all of their outlined hypotheses affect selfie-posting intention on social networks. Another work [52] extracts features from each images shared by users on social media platforms and shows that it is easy to distinguish between photos posted by healthy users and depressed ones. The results of this work demonstrate that healthy people will share photographs with higher hue value in comparison to depressed users who tend to share grayer images with lower brightness color. These results show that it is possible to detect depression through visual social media data. These findings confirm the fact that social media data could be used for mental health-related research.

Combined-Based Findings

A study by Andalibi et al. [3] find that Instagram users are aware of their audience. This is notable by observing how users address the audiences' concerns in captions of their posts. They find that images posted with a heavy amount of captions are related to support seeking and positive expression. Their results also show that specific hashtags on Instagram are used not only as semantic markers, but also as one way of categorizing content for the public. Similarly, Manikonda and De Choudhury [40] found that users used images to express their feelings such as emotional distress, and helplessness. Users' posts can be further used to understand how vulnerable and socially isolated they are. Results also show that images with various visual cues contribute to how users express themselves on Instagram. The authors finally conclude that Instagram is one of the mediums for users with mental health problems to seek help and also receive psychosocial support that they need. Another work by Burke et al. [10] demonstrates that direct communication affects users positively by making the user to bond with other users. This can further help users to feel less lonely. Their results also show that users with lower umber of interactions with others, tend to become more observer of other peoples' lives. In another work, Peng et al. [48] study the effect of owning a pet on personnel happiness by investigating posted images on social media. They compare users' happiness scores and find that pet owners were slightly happier than people who do not own pets. These results show the effectiveness of social media data in understanding users' behaviors and mental health related issues. In the next section, we summarize our findings and discuss potential future direction to expand research in this field.

5   Discussion and Future Directions

This chapter presents an overview of mental health related work that use social media data and machine learning in their studies. We discuss three key points focusing on mental health studies using social media data, namely, data, approaches, and findings. In the rest of this chapter, we provide a brief discussion followed by future work.

5.1 Discussion

Social media can affect users in many different ways. The main concern is if social media is beneficial to overcoming mental health problems among users. In [61], Facebook was rated as negative when it comes to cyber-bullying and bad sleeping patterns for users. But, when it comes to social support and building online communities, Facebook does help and was rated positively. Hence, it is important to make sure that social media be used in a good manner that can benefit users. Some significant effects of social media contributing to mental health well-being include: (1) social media can reduce stress in users through active communication with other users [13] and also provide information in capturing individual's present state of mind [43]; (2) social media is a popular channel for users to seek help and share information on the stigmatized issue. The anonymity of social media gives freedom to the user to express their feeling and might also improve in the stigmatized topic discussion [20]. (3) The use of social media with active communication may lead to improvement in the capability to share and understand others' feelings [10]. A study by Grieve et al. [31] indicates that Facebook connectedness may reduce depression and anxiety. Engaging with online communities can also give users the feeling of social appreciation through being understood [5].

On the other hand, some earlier studies point out the negative effects of social media to users. One finding is that users may have social isolation problems when they spend too much time on social media without having active communication with the online communities [16, 22]. Increasing social media use without interaction with other users may cause depression, anxiety, sleep problems, eating disorder, and suicide risk. Primack et al. [49] report that in terms of subjective social isolation or perceived social isolation (PSI), increased time spent on social media can result in decreased traditional social experience, thus, increased social isolation, and exacerbate the feeling of exclusion.

Figure 4 shows different types of mental illnesses and related work using social media data and analysis. Existing studies are summarized in categories of mental health issues: (1) mood disorder [22, 43, 45, 52], (2) post-traumatic stress disorder (PTSD) [15, 33, 46], (3) anxiety [20, 27, 56], (4) psychotic disorder [41], (5) eating disorder [11], (6) sexual and gender disorder [27], (7) suicidal behavior [16, 21, 37], and (8) attention deficit and hyperactivity disorder (ADHD) [14]. Mood disorder such as depression is studied in [15, 23, 60]. Depression is one of the mental health issues with high prevalence, receiving increasing attention lately. Limited work focuses on the psychotic disorder, eating disorder, and ADHD. These studies can also leverage social media data as it provides data about individuals' language and behavior [33]. Researchers use social media data to predict types of mental issues [11, 22, 23, 36]. In [43] the severity of users' mental issues is estimated using social media data. A similar finding is reported in [41]. Network information available in social media data is leveraged for studying mental health issues [22, 35].

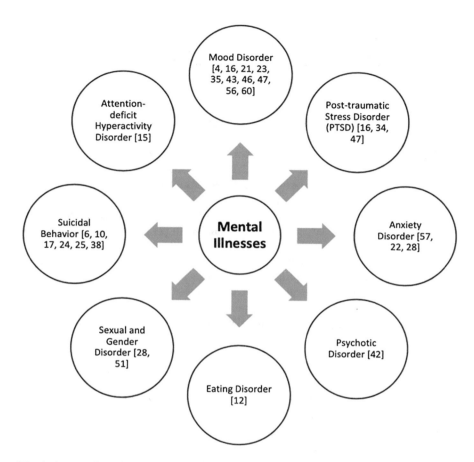

Fig. 4 An overview of different types of mental health issues and related work

5.2   Future Directions

Studying mental health issues using social media is challenging. Although a large body of work has emerged in recent years for investigating mental health issues using social media data, there are still open challenges for further investigation. Some potential research directions are suggested below:

– The increasing popularity of social media allows users to participate in online activities such as creating online profiles, interacting with other people, express-ing opinions and emotions, sharing posts and various personal information. User-generated data on these platforms is rich in content and could reveal information regarding users' mental health situation. However, little attention has been paid on collecting the proper amount of user-information specifically on mental health [16, 43]. One future direction is to collect a proper amount of labeled user-data as a benchmark which requires cooperation between psychologists and computer scientist [42]. This data can include users' behavioral information collected from social media platforms as well as their mental health condition information provided by experts. Preparing such data gives opportunities to both computer scientists and psychologists to benefit from a tremendous amount of data generated in social media platforms to better understand mental health issues and propose solutions to solve them.

– User-generated social media data is heterogeneous and consists of different aspects such as text, image, and link data. Most of the existing work investigates the mental health problem issues by just incorporating one aspect of social media data. For example, textual information is used in [2, 15, 16, 20, 22, 27, 43, 45, 55], image information is exploited [3, 36, 52], and link data in [22, 35] to understand how user-generated information is correlated with people's mental health concerns. One potential research direction is to examine how different combinations of heterogeneous social media data (e.g., a combination of image and link data, combination of textual and link data, etc.) can be utilized to better understand people's behavior and mental health issues concern. Another future direction is to explore how findings from each aspect of social media data are different from each other, e.g., results w.r.t. textual data in comparison the findings w.r.t. link data.

– Most existing work utilizes either human-computer interaction techniques or data mining related techniques. For example, interview and surveys are used to help further study mental health related issues in social media [10, 20, 22, 36, 45, 52]. Statistical and computational techniques are leveraged to understand users' behavior w.r.t. mental health issues [2, 35, 43, 55]. However, research can be furthered to exploit both techniques to understand mental health issues in social media [22, 45, 52] and to develop both human-computer interaction and computational techniques specialized for understanding mental health issues for social media data.

– This chapter shows how different mental health issues have been studied using social media data. Figure 4 represents different categories of mental health related issues using social media data. More mental health issues can be studied such as psychotic disorder, eating disorder, sexual and gender disorder.

Social media data can benefit mental health studies. The existing work shows that it is possible to study mental health by leveraging the large-scale social media data in understanding and analyzing mental health problems and employing machine learning algorithms to understand, measure, and predict mental health problems. More research on social media analysis using machine learning will help advance this important emerging field via multidisciplinary collaboration, research and development.

Acknowledgements We would like to thank all members of Data Mining Machine Learning Research Lab (DMML) at Arizona State University (ASU) for their constant support and feedback for this work. Special thanks to our lab members, Jundong Li, Matthew Davis, and Alex Nou for their detailed feedback on the earlier versions of this chapter. This work, in part, is supported by the Ministry of Higher Education Malaysia and University Malaysia Pahang (UMP).

References

1. I. Ajzen, The theory of planned behavior. Organ. Behav. Hum. Decis. Process. **50**(2), 179–211 (1991)
2. S. Amir, G. Coppersmith, P. Carvalho, M.J. Silva, B.C. Wallace, Quantifying mental health from social media with neural user embeddings (2017). Preprint. arXiv:1705.00335
3. N. Andalibi, P. Ozturk, A. Forte, Depression-related imagery on instagram, in *Proceedings of the 18th ACM Conference Companion on Computer Supported Cooperative Work & Social Computing* (ACM, New York, 2015), pp. 231–234
4. R. Astudillo, S. Amir, W. Ling, M. Silva, I. Trancoso, Learning word representations from scarce and noisy data with embedding subspaces, in *Proceedings of the 53rd Annual Meeting of the Association for Computational Linguistics and the 7th International Joint Conference on Natural Language Processing (Volume 1: Long Papers)*, vol. 1 (2015), pp. 1074–1084
5. D. Baker, S. Fortune, Understanding self-harm and suicide websites: a qualitative interview study of young adult website users. Crisis **29**(3), 118–122 (2008)
6. H. Bay, T. Tuytelaars, L. Van Gool, Surf: Speeded up robust features, in *European Conference on Computer Vision* (Springer, Berlin, 2006), pp. 404–417
7. D.M. Blei, A.Y. Ng, M.I. Jordan, Latent Dirichlet allocation. J. Mach. Learn. Res. **3**(Jan), 993–1022 (2003)
8. I. Bourgeault, R. Dingwall, R. De Vries, *The SAGE Handbook of Qualitative Methods in Health Research* (SAGE, London, 2010)
9. S.R. Braithwaite, C. Giraud-Carrier, J. West, M.D. Barnes, C.L. Hanson, Validating machine learning algorithms for twitter data against established measures of suicidality. JMIR Mental Health **3**(2), e21 (2016)
10. M. Burke, C. Marlow, T. Lento, Social network activity and social well-being, in *Proceedings of the SIGCHI Conference on Human Factors in Computing Systems* (ACM, New York, 2010), pp. 1909–1912
11. S. Chancellor, Z. Lin, E.L. Goodman, S. Zerwas, M. De Choudhury, Quantifying and predicting mental illness severity in online pro-eating disorder communities, in *Proceedings of the 19th ACM Conference on Computer-Supported Cooperative Work & Social Computing* (ACM, New York, 2016), pp. 1171–1184
12. W. Chris, Feeling overwhelmed by academia? You are not alone. Nature **557**, 129 (2018)
13. A. Cohan, S. Young, A. Yates, N. Goharian, Triaging content severity in online mental health forums. J. Assoc. Inf. Sci. Technol. **68**(11), 2675–2689 (2017)
14. G. Coppersmith, M. Dredze, C. Harman, K. Hollingshead, From ADHD to SAD: analyzing the language of mental health on Twitter through self-reported diagnoses, in *Proceedings of the 2nd Workshop on Computational Linguistics and Clinical Psychology: From Linguistic Signal to Clinical Reality* (2015), pp. 1–10
15. G. Coppersmith, M. Dredze, C. Harman, K. Hollingshead, M. Mitchell, Clpsych 2015 shared task: depression and PTSD on twitter, in *Proceedings of the 2nd Workshop on Computational Linguistics and Clinical Psychology: From Linguistic Signal to Clinical Reality* (2015), pp. 31–39
16. G. Coppersmith, R. Leary, E. Whyne, T. Wood, Quantifying suicidal ideation via language usage on social media, in *Joint Statistics Meetings Proceedings, Statistical Computing Section, JSM* (2015)

17. T. Cox, A. Griffiths, E. Rial-Gonzalez, Research on work related stress. Office for official publications of the European communities, Luxembourg, 2000
18. K. Danna, R.W. Griffin, Health and well-being in the workplace: a review and synthesis of the literature. J. Manag. **25**(3), 357–384 (1999)
19. H. Daumé III, Frustratingly easy domain adaptation (2009). Preprint. arXiv:0907.1815
20. M. De Choudhury, S. De, Mental health discourse on reddit: self-disclosure, social support, and anonymity, in *ICWSM* (2014)
21. M. De Choudhury, E. Kıcıman, The language of social support in social media and its effect on suicidal ideation risk, in *Proceedings of the International AAAI Conference on Weblogs and Social Media*, vol. 2017 (NIH Public Access, 2017), p. 32
22. M. De Choudhury, M. Gamon, S. Counts, E. Horvitz, Predicting depression via social media, in *ICWSM*, vol. 13 (2013), pp. 1–10
23. M. De Choudhury, S. Counts, E.J. Horvitz, A. Hoff, Characterizing and predicting postpartum depression from shared facebook data, in *Proceedings of the 17th ACM Conference on Computer Supported Cooperative Work & Social Computing* (ACM, New York, 2014), pp. 626–638
24. M. De Choudhury, E. Kiciman, M. Dredze, G. Coppersmith, M. Kumar, Discovering shifts to suicidal ideation from mental health content in social media, in *Proceedings of the 2016 CHI Conference on Human Factors in Computing Systems* (ACM, New York, 2016), pp. 2098–2110
25. N.H. El-Ghoroury, D.I. Galper, A. Sawaqdeh, L.F. Bufka, Stress, coping, and barriers to wellness among psychology graduate students. Train. Educ. Prof. Psychol. **6**(2), 122 (2012)
26. N.B. Ellison, C. Steinfield, C. Lampe, The benefits of facebook "friends" social capital and college students' use of online social network sites. J. Comput.-Mediat. Commun. **12**(4), 1143–1168 (2007)
27. M. ElSherief, E.M. Belding, D. Nguyen, # notokay: Understanding gender-based violence in social media, in *ICWSM* (2017), pp. 52–61
28. T.M. Evans, L. Bira, J. Beltran-Gastelum, L.T. Weiss, N. Vanderford, Mental health crisis in graduate education: the data and intervention strategies. FASEB J. **31**(1 Supplement), 750–757 (2017)
29. T.M. Evans, L. Bira, J.B. Gastelum, L.T. Weiss, N.L. Vanderford, Evidence for a mental health crisis in graduate education. Nat. Biotechnol. **36**(3), 282 (2018)
30. B. Gentile, J.D. Miller, B.J. Hoffman, D.E. Reidy, A. Zeichner, W.K. Campbell, A test of two brief measures of grandiose narcissism: the narcissistic personality inventory-13 and the narcissistic personality inventory-16. Psychol. Assess. **25**(4), 1120 (2013)
31. R. Grieve, M. Indian, K. Witteveen, G.A. Tolan, J. Marrington, Face-to-face or facebook: can social connectedness be derived online? Comput. Hum. Behav. **29**(3), 604–609 (2013)
32. E. Gunther, W. Jeremy, Using the internet for surveys and health research. J. Med. Internet Res. **4**(2), E13 (2002)
33. G. Harman, M.H. Dredze, Measuring post traumatic stress disorder in twitter, in *ICWSM* (2014)
34. C.M. Homan, N. Lu, X. Tu, M.C. Lytle, V. Silenzio, Social structure and depression in TrevorSpace, in *Proceedings of the 17th ACM Conference on Computer Supported Cooperative Work & Social Computing* (ACM, New York, 2014), pp. 615–625
35. I. Kawachi, L.F. Berkman, Social ties and mental health. J. Urban Health **78**(3), 458–467 (2001)
36. E. Kim, J.-A. Lee, Y. Sung, S.M. Choi, Predicting selfie-posting behavior on social networking sites: an extension of theory of planned behavior. Comput. Hum. Behav. **62**, 116–123 (2016)
37. M. Kumar, M. Dredze, G. Coppersmith, M. De Choudhury, Detecting changes in suicide content manifested in social media following celebrity suicides, in *Proceedings of the 26th ACM Conference on Hypertext & Social Media* (ACM, New York, 2015), pp. 85–94
38. K. Levecque, F. Anseel, A. De Beuckelaer, J. Van der Heyden, L. Gisle, Work organization and mental health problems in phd students. Res. Pol. **46**(4), 868–879 (2017)
39. Z. Lin, N. Salehi, B. Yao, Y. Chen, M.S. Bernstein, Better when it was smaller? Community content and behavior after massive growth, in *ICWSM* (2017), pp. 132–141

40. L. Manikonda, M. De Choudhury, Modeling and understanding visual attributes of mental health disclosures in social media, in *Proceedings of the 2017 CHI Conference on Human Factors in Computing Systems* (ACM, New York, 2017), pp. 170–181
41. M. Mitchell, K. Hollingshead, G. Coppersmith, Quantifying the language of schizophrenia in social media, in *Proceedings of the 2nd Workshop on Computational Linguistics and Clinical Psychology: From Linguistic Signal to Clinical Reality* (2015), pp. 11–20
42. F. Morstatter, J. Pfeffer, H. Liu, K.M. Carley, Is the sample good enough? Comparing data from twitter's streaming API with twitter's firehose, in *Seventh International AAAI Conference on Weblogs and Social Media* (2013)
43. M. Nadeem, Identifying depression on twitter (2016). Preprint. arXiv:1607.07384
44. S.J. Pan, Q. Yang, A survey on transfer learning. IEEE Trans. Knowl. Data Eng. **22**(10), 1345–1359 (2010)
45. M. Park, C. Cha, M. Cha, Depressive moods of users portrayed in twitter, in *Proceedings of the ACM SIGKDD Workshop on Healthcare Informatics (HI-KDD)*, vol. 2012 (ACM, New York, 2012), pp. 1–8
46. T. Pedersen. Screening twitter users for depression and ptsd with lexical decision lists, in *Proceedings of the 2nd Workshop on Computational Linguistics and Clinical Psychology: From Linguistic Signal to Clinical Reality* (2015), pp. 46–53
47. F. Pedregosa, G. Varoquaux, A. Gramfort, V. Michel, B. Thirion, O. Grisel, M. Blondel, P. Prettenhofer, R. Weiss, V. Dubourg et al., Scikit-learn: machine learning in python. J. Mach. Learn. Res. **12**(Oct), 2825–2830 (2011)
48. X. Peng, L.-K. Chi, J. Luo, The effect of pets on happiness: a large-scale multi-factor analysis using social multimedia. ACM Trans. Intell. Syst. Technol. (TIST) **9**(5), 1–15 (2018)
49. B.A Primack, A. Shensa, J.E. Sidani, E.O. Whaite, L. yi Lin, D. Rosen, J.B. Colditz, A. Radovic, E. Miller, Social media use and perceived social isolation among young adults in the US. Am. J. Prev. Med. **53**(1), 1–8 (2017)
50. H. Purohit, T. Banerjee, A. Hampton, V.L. Shalin, N. Bhandutia, A.P. Sheth, Gender-based violence in 140 characters or fewer: a# bigdata case study of twitter (2015). Preprint. arXiv:1503.02086
51. L.S. Radloff, The CES-D scale: a self-report depression scale for research in the general population. Appl. Psychol. Meas. **1**(3), 385–401 (1977)
52. A.G. Reece, C.M. Danforth, Instagram photos reveal predictive markers of depression. EPJ Data Sci. **6**(1), 15 (2017)
53. S. Rude, E.-M. Gortner, J. Pennebaker, Language use of depressed and depression-vulnerable college students. Cognit. Emot. **18**(8), 1121–1133 (2004)
54. D.W. Russell, UCLA loneliness scale (version 3): reliability, validity, and factor structure. J. Pers. Assess. **66**(1), 20–40 (1996)
55. K. Saha, M. De Choudhury, Modeling stress with social media around incidents of gun violence on college campuses. Proc. ACM Hum. Comput. Interact. **1**(CSCW), 1–27 (2017)
56. B. Shickel, M. Heesacker, S. Benton, P. Rashidi, Hashtag healthcare: from tweets to mental health journals using deep transfer learning (2017). Preprint. arXiv:1708.01372
57. Social media can be bad for youth mental health, but there are ways it can help. https://theconversation.com/social-media-can-be-bad-for-youth-mental-health-but-there-are-ways-it-can-help-87613. Accessed 15 Mar 2018
58. K.-J. Stol, B. Fitzgerald, Two's company, three's a crowd: a case study of crowdsourcing software development, in *Proceedings of the 36th International Conference on Software Engineering* (ACM, New York, 2014), pp. 187–198
59. A.H. De Lange, T.W. Taris, M.A.J. Kompier, I.L.D. Houtman, P.M. Bongers, Work characteristics and psychological well-being. Testing normal, reversed and reciprocal relationships within the 4-wave smash study. Work Stress **18**(2), 149–166 (2004)
60. S. Tsugawa, Y. Kikuchi, F. Kishino, K. Nakajima, Y. Itoh, H. Ohsaki, Recognizing depression from twitter activity, in *Proceedings of the 33rd Annual ACM Conference on Human Factors in Computing Systems* (ACM, New York, 2015), pp. 3187–3196

61. UK News, Instagram rated worst social network for mental health (2017). http://www.theweek. co.uk/84799/instagram-rated-worst-social-network-for-mental-health. Accessed 28 Mar 2018
62. N.A. Van House, M. Davis, The social life of cameraphone images, in *Proceedings of the Pervasive Image Capture and Sharing: New Social Practices and Implications for Technology Workshop (PICS 2005) at the Seventh International Conference on Ubiquitous Computing (UbiComp 2005)*. Citeseer (2005)
63. M. Worring, Lecture notes: Multimedia information systems (2015). http://citeseerx.ist.psu. edu/viewdoc/download?doi=10.1.1.103.6399rep=rep1type=pdf

Automated Text Analysis for Intelligence Purposes: A Psychological Operations Case Study

Stefan Varga (iD), Joel Brynielsson (iD), Andreas Horndahl (iD), and Magnus Rosell (iD)

Abstract With the availability of an abundance of data through the Internet, the premises to solve some intelligence analysis tasks have changed for the better. The study presented herein sets out to examine whether and how a data-driven approach can contribute to solve intelligence tasks. During a full day observational study, an ordinary military intelligence unit was divided into two uniform teams. Each team was independently asked to solve the same realistic intelligence analysis task. Both teams were allowed to use their ordinary set of tools, but in addition one team was also given access to a novel text analysis prototype tool specifically designed to support data-driven intelligence analysis of social media data. The results, obtained from the case study with a high ecological validity, suggest that the prototype tool provided valuable insights by bringing forth information from a more diverse set of sources, specifically from private citizens that would not have been easily discovered otherwise. Also, regardless of its objective contribution, the capabilities and the usage of the tool were embraced and subjectively perceived as useful by all involved analysts.

Keywords Data-driven analysis · Text analysis · Social media · Intelligence · Psychological operations · Ecological validity

S. Varga
KTH Royal Institute of Technology, Stockholm, Sweden

Swedish Armed Forces Headquarters, Stockholm, Sweden
e-mail: svarga@kth.se

J. Brynielsson (✉)
KTH Royal Institute of Technology, Stockholm, Sweden

FOI Swedish Defence Research Agency, Stockholm, Sweden
e-mail: joel@kth.se

A. Horndahl · M. Rosell
FOI Swedish Defence Research Agency, Stockholm, Sweden
e-mail: andreas.horndahl@foi.se; magnus.rosell@foi.se

1 Introduction

Vast amounts of easily accessible data on the Internet is generated every day by news outlets, individuals, and other information sources. The traffic volumes that produce this sea of data, too big for any human or group of humans to process without help, are predicted to increase even more in the future [13]. In other words, *information overload*, which simply put is about receiving too much information (to handle) [20], sometimes impairs the human ability to make use of available information. As the right pieces of information have the potential to be of value to some individuals or groups, it is desirable to possess a capability that takes advantage of the possibilities of online data to the largest extent possible.

A set of tools that enable users to sift through huge amounts of data in search of sought after information that is relevant to them, is software in the form of data mining and other analytical tools. It is generally thought that the end outcomes for various stakeholders will improve when such tools and techniques are employed in a systematic fashion. Even if that assumption seems highly plausible, there appears to be limited research that either confirms or proves such a hypothesis wrong. On the contrary, the (positive) value of text mining and similar techniques are often taken as a given fact in the literature.

This work seeks to examine the actual contribution of the use of a text mining tool for a real life intelligence task. In the literature much attention has been paid to the investigation of matters related to the *usability* of software in terms of design and function [44]. When it comes to the actual *usefulness* of software in relation to some task, considerably less scholarly articles are available. Furthermore, it is not always obvious that computer software contribute to the productivity of organizations at all [35, 45]. It is obviously of value to know if a piece of software actually contributes to overall organizational goals or not. In this chapter the usefulness of a piece of text analysis software for the purpose of intelligence analysis is examined.

1.1 The Intelligence Field

Because the context of this case study is set in an intelligence analysis setting, some words about the intelligence field is given to enlighten the reader.

It seems that intelligence is not consistently defined in the literature, as many articles within the field begin with quite extensive discussions about basic definitions [6]. In the national level context, however, Bimfort already in 1958 [7] suggested that intelligence is about collecting and processing information about foreign countries and their agents, that is needed by a government for its foreign policy and national security goals. This definition will stand for the purpose of this chapter as well.

According to U.S. doctrine Joint Publication 2-01 [51], which has similar definitions as other countries, the objective of joint intelligence operations in a military context is to provide accurate and timely intelligence to commanders that gives them an understanding of the operational environment with, in particular, regard to adversary forces, capabilities, and intentions. The goal of intelligence analysis is therefore to produce accurate assessments of the current state of affairs, as well as sufficiently good estimates about the future that can be of use to various decision-makers [14].

Although this mission is simple enough in theory, (good) analytical work is not trivial. A number of uncertainties affect the quality of the final analysis, and the process itself is not straightforward as it typically involves a mixture of imaginative and critical reasoning [8]. Sometimes analysts are in possession of a piece of information, and in search for missing pieces to fit into a hypothesis. At the same time the analysts may also search for supportive evidence for multiple hypotheses [8]. In addition, most intelligence processes strive to collect information from several mutually independent sources in order to, hopefully, reduce uncertainty. Open source information that is readily available and easily accessible is certainly often used in this respect.

In a military intelligence context the scope of the analytical interest for different "situations" as discussed above, varies with the hierarchical levels of war, e.g., the tactical, the operational, and the strategic level, which highlights diverse aspects of awareness. Although the exact meaning and classification of the different levels vary between nations and organizations, the existence of multiple hierarchical levels is uncontroversial. In the following, Swedish doctrine is used to exemplify. The differences and boundaries between the levels have diminished over time due to the dynamics and complexities of modern conflicts. The same pieces of information can sometimes answer to the intelligence requirements of multiple levels. The greatest distinguishing factor between the levels is the time perspective [48]. The goal of the tactical level intelligence function is to support the individual (military) unit in its planning and mission execution. The tactical level deals with a limited geographical region on the battlefield and the character of the questions that need answers are concrete. The relevant timeframe for these activities is the immediate; it ranges from a day to a week, and sometimes from an hour to days [48]. The goal of the operational level intelligence function, on the other hand, is to support ongoing or planned (military) operations. Here the area of operation (theatre) is the region of interest. The questions are diverse, both concrete and abstract [24], and the timeframe is the intermediate; it typically ranges from a week to months [48]. At the strategic level the goal is to answer questions that may be abstract and cover a diversity of different topics with unclear relations, and to produce intelligence in the form of estimates that can be used to, e.g., create policies and military plans, and to inform about security measures on a national level. The timeframe for strategic intelligence ranges from months to several years [10]. The strategic level, thus, concerns and anticipates events of far-reaching political, diplomatic, social, economic, and military significance, that often revolve around the questions of war, peace, and stability [37].

With the wide variety of requirements as indicated above, a national level intelligence agency needs to have different types of intelligence sources at its disposal. A commonly accepted taxonomy divide some of the source types into open source intelligence (OSINT), human intelligence (HUMINT), measurements and signatures intelligence (MASINT), signals intelligence (SIGINT), and imagery intelligence (IMINT) [14, p. 104]. With the rapid growth of the Internet and the availability of data, OSINT has grown to become an important collection discipline for intelligence purposes; not only for government intelligence functions, but also for civilian use [46]. OSINT, more specifically, is intelligence that fulfills specific intelligence requirements and is produced from publicly available information from both traditional media and web-based sources, that is, information that anyone can lawfully obtain by request, purchase, or observation [51]. Glassman and Kang [22] characterize the baseline work methods for OSINT work to be (1) the search for relevant information, (2) the organization of the information, and (3) the differentiation of it, with the goal of transforming (converting, translating, and formatting) text, graphics, sound, and motion video in response to users' intelligence requirements [51].

Heuer [27] suggests that there are in general two fundamental but different methods to address intelligence analysis problems: the conceptually driven approach, and the data-driven approach. The conceptually driven approach requires the presence of an analytical schema, e.g., some model, and the results of the analysis are directly correlated to the availability and the quality of data that is fed into the model. For data-driven analysis, on the other hand, there may not be a well-developed analytical model available, but rather an abundance of available data. In this latter case the challenges are not mainly about acquiring the data, but rather to find and select the relevant pieces of information that can be used to form sensible hypotheses for the intelligence problem at hand. A potential source of error when using the conceptual approach, is that research has shown that long-standing general beliefs and preconceptions by analysts affect the results of assessments even if there is available data that support other outcomes [33].

To this respect the availability of vast amounts of data that the Internet provides, has changed the work for intelligence analysts [18], and is a good match for the requirements of data-driven analysis. In particular, it has been found that social media posts provide relevant data that may be exploited for intelligence analysis. Several types of analyses can be made based on social media data, e.g., text analysis, social network analysis, and trend analysis [47].

1.2   Research Questions and Outline

The overall purpose of the research presented in this chapter is to investigate how automated text analysis can contribute to solve ordinary intelligence analysis tasks. To do this, a case study that sought to answer the following two research questions was performed:

- Is PhraseBrowser, a specific instantiation of a text analysis tool, perceived as useful for solving typical analytical tasks?
- Does the use of the text analysis tool improve the quality of a typical analytical deliverable?

To address the research questions, a research design that asked analysts to make an intelligence assessment of a psychological operations case was created. The case of the poisoning of the former Russian intelligence officer Sergei Skripal and his daughter Yulia in Great Britain was chosen. The rationale for choosing this incident was that it was well covered by multiple media outlets of different types as well as other data producing sources. The topic was reasonably well within the ordinary field of interest for the studied subjects, the scope of the intelligence analysis requirement was fairly realistic, and some amount of different narratives and counter-narratives, e.g., obfuscation, could be expected to be launched and spread to flourish on the Internet.

The remainder of this chapter is structured as follows. Section 2 provides an overview of the area of psychological operations, and related work. Then, in Sect. 3, a description of the text analysis prototype tool PhraseBrowser follows. Next, Sect. 4 describes the undertaken methodology covering the research design, the observational study setup, and the execution of the study. In Sect. 5 the results are presented. Then, the findings as well as issues of validity are discussed in Sect. 6. Finally, the conclusions are presented in Sect. 7.

2 Background

After having introduced the intelligence *field* above, this section describes the framework for intelligence *analysis*, psychological and influence operations, and lists some related work.

2.1 Intelligence Analysis

The basic process for producing intelligence is about collecting information, making analyses, and then creating an intelligence product that can form the basis for decision-making. It is primarily a cognitive process, an activity that takes place in the analyst's own head. It is not always clear exactly how the assessment goes and how it ought to be done in different situations, but on a methodological level there are strong links between the intelligence profession and that of scientific work: intelligence work is largely about formulating hypotheses, and examining and falsifying these hypotheses if possible. But there are also large differences relative to scientific work in that intelligence work has a connection to operational work and the need to deliver forecasts within a given timeframe based on currently available information, regardless of whether it is judged to be of sufficient quality or not.

Rather than having the character of a "secret science" *in itself*, the characteristic features of intelligence work can thus be said to be about the (indeed often secret) *application* of scientific methods and approaches on information and intelligence questions that are operational and strategic, rather than scientific, in nature. In an analysis of the intelligence subject, and in an attempt to go from established practice to a more structural approach of it, Agrell and Treverton [1, p. 279] state this in the following terms:

> Intelligence analysis has the potential to become an applied science. Its purpose would be managing the uncertainty in assessments of threats and possibilities based on incomplete, unreliable, or uncertain data in a context in which demand requires those assessments irrespective of the limitations. Defined in these terms, intelligence analysis stands out as a genuine cross-disciplinary science in-being, with a theoretical basis and a set of methods not limited to any single subject matter or field of analysis but rather adapted to every specific application.

As noted, management of information and its related uncertainty play a central role, and the means to measure precision, quality, and utility—so-called information awareness [5]—is crucial.

2.2 Psychological Operations

The *root causes* for armed conflicts, or indeed any controversies between humans, may be of different kinds, e.g., ideological differences, competition for scarce resources, etc., but the *end goal* in conflicts is always to impose the will of one party on the other(s) in one way or the other, as expressed by, e.g., the ancient Chinese war theorist Sun Tzu [49], as well as the highly influential Preussian war theorist Carl von Clausewitz [54]. But there are other methods to affect the will of an opponent than by using physical force: psychological and influence operations are ways to achieve this.

Psychological operations, Psyops, within the context of military operations is a part of offensive information operations that aim to influence perceptions, attitudes and ultimately change the behavior of foreign approved target audiences [50]. Another name for Psyops that is sometimes used, is military information support operations, MISO [52]. Such activities are sometimes seen as a key enabler in a military commander's campaign plan [4, 50], and they can be conducted with both short-term and long-term goals, according to doctrine. The overarching information operations field also involves other activities such as civil affairs, computer network attack, deception, destruction, electronic warfare, operations security, and public affairs [4]. Psyops and related activities, however, are not only part of U.S. and Nato doctrine, but integral parts of Russian, Chinese, and other countries' national security or military doctrines as well [4]. A related term is influence operations, which consists of similar types of activities, which are not necessarily conducted in conjunction with military operations. Influence operations primarily consist of non-kinetic, communications-related, and informational activities that aim to affect cognitive, psychological, motivational, ideational, ideological, and moral characteristics of a target audience [36]. Influence operations can be carried out by government organizations other than the military, as well as by civilian information outlets. Furthermore, operations can be conducted openly (with a named source), covertly, and clandestine.

The boundaries between roles, mandates, and who does what, as well as between military and civilian organizations, with regard to influence operations is not always clear-cut [36]. For the sake of simplicity, only the term Psyops is used in this text. The approved targets for operations can be individual persons, groups and networks, adversary leadership coalitions, and the (mass) public [36]. On the defensive side of Psyops, a possible task would be to determine if oneself, or some other target audience, is on the receiving end of adversarial operations. Such operations could aim to change the perceptions/views, clog the perceptions, or delegitimize credible news outlets [39]. To accomplish such an analysis, Nato doctrine [39], which is used to highlight Psyops principles here, prescribes that a detailed examination of source, content, audience, media, and effect (SCAME) is carried out. The SCAME template can also be used in the planning process for offensive operations.

2.3   Related Work

There seems to be a general consensus about the perceived usefulness of data mining tools, in that they can help find useful information in many instances. Examples of precisely how such tools contribute to the information gathering efforts, however, and more to the point, the quality improvement of the end products, are harder to find. A study commissioned by the British Government [38], however, concluded that text mining contributes with several benefits in the context of academic research. The study found one improvement to be overall increased researcher efficiency and research quality. Mining was further attributed to bring about the ability to "unlock hidden information", i.e., insights about underlying non-obvious connections between texts, as well as the capacity to develop new knowledge and the ability to explore "new horizons" [38, p. 19]. The conclusions were based on results from several case studies.

Pal [41] listed application areas where data mining tools were of great use both within the public and the commercial sector. For the public sector, fields such as, for example, scientific enquiry and research analysis, criminal investigation and homeland security, via health insurance and healthcare applications, were found to benefit from text mining. Within the purely commercial sector, areas such as customer segmentation and targeted marketing, finance, etc., were mentioned.

Several scholars have noted that publicly available data on the Internet can be systematically used for intelligence purposes [18, 40]. It has been observed that some of the automated data processing techniques used for systematic processing of corporate data, i.e., business intelligence, have migrated into security intelligence [10], and that the further development of these techniques has the potential to provide a competitive edge [12, 31, 42]. There are, however, challenges as of how to find and extract relevant and meaningful pieces of information relative to the task at hand.

Automated text analysis using methods and tools from the field of natural language processing has been proposed as a way to off-load some of the selective and interpretative work from human intelligence analysts. With regard to more closely related government and military intelligence tasks, Guo et al. [23] did work on entity extraction from human-generated tactical reports to support intelligence analysis. They extracted entities such as organizations, locations, persons, etc., with promising results. Razavi et al. [43] sought to extract information about risks in maritime operations.

Other examples with applications from the commercial sector include work by He et al. [25] that explored how the use of text mining was useful for companies in the pizza industry. They concluded that text mining of social media adds useful pieces of information, for example, in companies' quests to understand the pizza market. Alex et al. [2] were able to show an increase in efficiency, e.g., by the reduction of work time, in a biomedical data system scenario where natural language processing technology was used for curation.

3   PhraseBrowser

This section serves to describe the scope and function of the prototype tool PhraseBrowser. PhraseBrowser is continuously developed at the Swedish Defence Research Agency. It is one of several prototypes in a framework, aiming to highlight the possibilities given by web and text analysis tools to analysts. The development process of PhraseBrowser, to this respect, provides opportunities for researchers and practitioners to engage in mutually beneficial discussions.

3.1   Overview

PhraseBrowser is a text analysis prototype tool designed to support analytical work by processing Twitter data. The idea is to provide the user/analyst with several perspectives of the collected data through predefined themes. The different perspectives provide an overview of the data that may guide an analyst to select interesting subtopics and drill down further to find specific contents of interest. Hence, the prototype tool may be useful both for monitoring a subject or area of interest, and for conducting research to answer specific questions. Although the prototype tool comes with predefined themes, it is a relatively easy and straightforward process to quickly add simple first versions of new themes. This versatility makes the tool relevant for analytical work that concerns a wide variety of topical issues as well as for time-sensitive tasks.

3.2 Phrases

PhraseBrowser presents phrases to the user sorted by so-called phrase types. A phrase is defined as a sequence of one or more words, and each phrase has a type. A few example phrases (and their phrase types) from the set of tweets used in the study are: "lab says" (phrase type: "General Phrase"), "Boris Johnson lied about Skripal" ("Explicit Untruth"), "UK" ("Location"), "Yulia" ("Person"), and "30 questions" ("Counted Thing"). See Table 1 for more examples of phrase types and phrases. At the time of the study there were more than 50 different phrase types of varying quality available to the analysts.

In the interface the user can choose a type, e.g., "Counted Things", leading to that a list of phrases of that type are presented along with statistics on how many tweets the phrases were used in. To read the texts containing any particular phrase, e.g., "30 questions", the user simply clicks on that phrase. Each piece of text works as a link to the original tweet on Twitter, where more context may be found by studying the content of corresponding accounts, etc.

The phrases may be identified using any automatic method. In the version of the tool used in the observational study presented herein, a third party library was used to identify entities (see Sect. 3.5), and the rule language described in Sect. 3.6 was used for all other phrase types.

3.3 Predefined Phrase Types and Filtering

In the current study the analysts worked with a set of predefined phrase types, and were not able to alter them or add new ones. Table 1 displays some of these phrase types, and in the following they are described a bit further:

"General Phrases" tries to capture any kind of content based on part of speech tags. This is an example of a phrase type that results in many phrases—perhaps too many. It would probably be useful to exchange or complement this phrase type with a machine learning method. For now this is the most general phrase type, primarily used to explore content without looking for any of the specific content that most of the other phrase types try to capture.

"Counted Things/Persons" is defined using other phrase types capturing counts, and at the same time "things" and/or "persons". One possible use of this phrase type is to look for differing numbers being given in some context. Sources may exaggerate the number of protesters at an event, for instance.

"Entities" such as "Person", "Location", and "Organization" are found by an entity detector (see Sect. 3.5). These entities are reused by several of the other phrase types, e.g., the "Counted Persons" phrase type mentioned above.

"Explicit Untruths" captures phrases that use any word in a long list of words explicitly related to deception, propaganda, misinformation, fake news, etc. The idea is that it is potentially interesting whenever someone writes that something is

Table 1 Examples of themes/types available in the PhraseBrowser tool along with phrases extracted from the data set used in the study

Theme/Type	Phrase	#
General Phrases	*All*	2,806,084
	lab says	9297
	used in Skripal poisoning	8842
	produced in Russia	5284
	disinformation campaign in Britain	4326
	was in US	4205
	Porton Down research laboratory	3401
	research laboratory has told Sky News	3373
	to Sergei Skripal	3122
	that Christopher Steele	2803
	of Yulia Skripal	2563
Counted Things/Persons	*All*	105,931
	60 Russian diplomats	1406
	30 questions	927
	two weeks	861
	14 simple questions	823
	hundred narratives	383
	two BBC colleagues	294
	20 European countries	280
	2800 Russian bots	232
	23 British diplomats	212
	two people poisoned	203
Entities	*All*	5,601,461
	Skripal	373,735
	Russia	113,772
	UK	107,982
	Yulia	97,560
	Salisbury	32,903
	Porton	25,029
	Novichok	22,236
	Putin	22,017
	OPCW	19,292
	Theresa May	14,136
Explicit untruths	*All*	62,579
	propaganda	4532
	Boris Johnson lied about Skripal	1641
	Moscow's lies	427
	Kremlin propaganda	417
	UK lies	204
	Theresa May is lying	155
	"Russia bot" narrative	153
	the Skripal narrative	142
	lying about the source of Novichok	111
	Downing Street spin-master	82

a lie, fake, misinformation, etc.: either the statement is true, or the person writing it has an agenda. . . Hence, these phrases often contain suspicious/interesting statements that can potentially be used as a starting point for further analysis. In Sect. 3.7 some simplified rule examples for "Explicit Untruths" are presented.

To drill down into the data, each phrase and phrase type may be used as a filter that narrows the search to only include tweets containing the chosen phrases. The resulting smaller data set can then be studied using the other available types and phrases. For instance, filtering the data using the phrase "Theresa May" and then looking at the phrase type "Explicit Untruths", one would only obtain the "Explicit Untruths" that co-occur with "Theresa May".

3.4   PhraseBrowser System Overview

Figure 1 shows an overview of the PhraseBrowser system. Data is downloaded in real time using the Twitter Streaming API and/or RSS feeds. The data is continuously processed by several analysis components, and stored in an Elasticsearch[1] database search engine. The system architecture and construction is scalable and adapted to parallelization, using Docker[2] for packaging of subsystems and Kafka[3] for distributing data between the subsystems.

The analysis phase contains several subsystems that process the data and adds information to it. The original data (including metadata, if any) is stored by the storage component along with the result of the analysis as (additional) metadata.

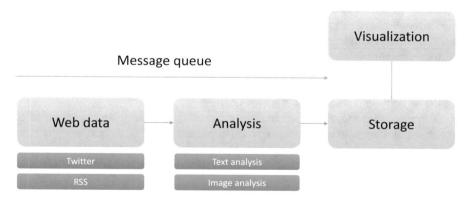

Fig. 1 PhraseBrowser system sketch. The prototype tool can handle Twitter and RSS data, and analyzes both text and images

[1] https://www.elastic.co/elasticsearch/.

[2] https://www.docker.com/.

[3] https://kafka.apache.org/.

By filtering on the metadata, different parts of the stored data can be retrieved and visualized in the user interface.

PhraseBrowser can be run in real time on a data stream. However, depending on the hardware used, the analysis may not keep up with the stream. In such cases PhraseBrowser continuously processes the latest tweet or RSS update, meaning that certain pieces of data may never be processed. In the present observational study a single ordinary PC was used to run the search query "Skripal,skripal" using the Twitter Streaming API, leading to that 5% of the tweets were discarded.

3.5   Text Processing

Figure 2 magnifies the part of the (text) analysis steps relevant to this study. For detecting most phrases, a rule-based approach [30] is applied. Each piece of text is first run through a natural language processing (NLP) library to divide it into sentences and tokens, lemmatize tokens and determine part of speech, and detect entities. The output from the NLP library is transformed into a simple specific text format representing each sentence in the input data. The formatted text is then sent to the rule language engine, resulting in a set of phrases for each sentence. The phrases are added to the metadata for the tweet/text as described in Sect. 3.4.

Through this construction, the rule language engine is separated from the NLP library and the only thing that has to be done to try a new library is to specify the transformation from the NLP library output format to the specific text format needed by the rule language engine. Transformations have been specified for a few different libraries. For the current work TwitIE [9], a GATE [15] pipeline specialized for microblog texts, was used.

The transformation into the previously mentioned specific text format includes turning the entities detected by the NLP library into phrases, with phrase types corresponding to the entity type. Hence, these phrases are originally detected by whatever method the library uses. For entities this is usually a combined method based on both dictionaries and machine learning.

The rule language, described further in Sect. 3.6, is flexible and constructed to make it possible to detect anything from single words using simple word lists to more complex phrases. It is language dependent in the sense that for any new language it is necessary to create a parallel text processing pipeline according to

Fig. 2 PhraseBrowser text processing. Each piece of text is processed by the NLP library, followed by a transformation of the NLP library output into a specific text format. Based on the formatted text, the rule language engine produces phrases for each sentence

Fig. 2. However, after an initial learning period, it is easy to quickly create basic phrase types for a new language. So far processing pipelines and rules for English and Swedish have been implemented.

3.6 PhraseBrowser Rule Language

The rule language can be compared to regular expressions, but on token (word) level rather than character level. The concept is inspired by many previous such pattern detection methods, like for instance Hearst's patterns for finding hyponymy relations [26, 32].

Each phrase type is defined using several rules. The rule language allows for references to other phrase types, making it possible to reuse solutions and create more sophisticated rules. The rule engine processes each sentence by applying the rules in order of complexity, making sure that a rule referring to another rule is applied after the rule it is referencing has already been applied.[4] Every result is temporarily kept in a data structure containing information about position in the sentence, so that results do not need to be found twice. This makes the rule engine efficient, and less likely to become the bottleneck of the larger system. However, with a huge set of rules and/or rules that are too generic and always find phrases, the rule engine could still become a problem.

During the text analysis the rule language is applied to a single sentence at a time. If larger text blocks were to be considered, i.e., several sentences are used, too many phrases would be selected. Hence, to capture information that is distributed over several sentences, other methods need to be used. Such methods were, however, not used in the study presented herein, and will therefore not be discussed further.

3.7 PhraseBrowser Rule Language Examples

Table 2 shows a few simple examples of rules written with the rule language. The left column contains rules for two different example phrase types called "violence" and "violence_in_location". The first phrase type is a word list that simply detects whenever the listed words appear, as in the example sentences in the right column. The rule language has several features that allow for creating more useful word lists, such as using the lemma instead of the actual token and using part of speech tags.

The phrase type "violence_in_location" contains two rules that reference the "violence" phrase type and the entity type "LOCATION". The two rules also allow

---

[4]Circular references are not allowed.

Table 2 Simple examples of the PhraseBrowser rule language

#	Phrase type and rules	Example text
	name:violence	
1	violence	London has seen a lot of **violence**.
2	fist fight	There is a **fist fight** in London.
	name:violence_in_location	
3	!a(LOCATION) !any[*] !a(violence)	**London has seen a lot of violence**.
4	!a(violence) !any[*] !a(LOCATION)	There is a **fist fight in London**.

The left column shows simple rules, with rule numbers for convenience. The phrase type "violence_in_location" reuses the phrase type "violence". The right column provides examples of applying the rules. The boldface part of the example sentences is what would be found by the rule in the same row

for any number (zero or more) of any kind of token in between.[5] The first rule in "violence_in_location" (row 3 in Table 2) therefore could be read as:

> a location entity, followed by zero or more appearances of any token, followed by either "violence" or "fist fight".

The second rule (row 4 in Table 2) can be interpreted analogously. In practice the "violence_in_location" rules would detect too many uninteresting phrases, as sentences may be long and the mentioning of a location does not necessarily relate to where the "violence" is taking place. To overcome this problem, the rule language has features for stopping phrases that contain certain tokens (or phrase types) between parts of the rules, and it also allows for requiring the presence or absence of certain tokens (or phrase types) within the current sentence.

In Table 3 some simplified examples of the rules for the "Explicit Untruths" phrase type are presented. The rules show the expressive power of the rule language: if the building blocks are well thought-out, the rules can become capable of detecting many different kinds of relevant phrases. It is often easier to split more complex phenomena into parts. The "Explicit Untruth" rules used in the study, for instance, are split into a few subtypes. The precise rule structure can be accomplished using linguistic insights, but more importantly it should be based on the data at hand and be useful for the analyst.

It is imperative that either the analyst is actively involved in the creation of the rules or works in a team with an expert, since otherwise the analyst may interpret the results erroneously. An analyst at the least needs to be made aware of the limitations of the rule language. Trial and error brings the analyst a long way in creating rules that find many relevant examples, though. The precision can become high enough, while the recall obviously cannot be guaranteed.

---

[5]It is possible to specify any number of repetitions, like for instance two to three tokens, as well as specifying that only tokens from other phrase types are allowed in the repetition.

Table 3 Simplified examples of the "Explicit Untruth" rules

#	Phrase type and rules	Example text
	name:u_basic	
1	lie	Boris Johnson **lied** about Skripal
2	propaganda	It's all Kremlin **propaganda**
3	spread rumor	John **spreads rumors** about Paul
4	misinformation	The fight against **misinformation**
	name:u_obj	
5	!a(PERSON)	**Boris Johnson** lied about **Skripal**
6	!a(LOCATION)	It's all **Kremlin** propaganda
	name:untruth	
7	!a(u_basic)	That's just **lies**
8	!a(u_obj) !a(u_basic)	It's all **Kremlin propaganda**
		Many **UK lies** today
9	!a(u_obj) 's !a(u_basic)	Have you heard **Moscow's lies**?
		We are used to **John's misinformation**
10	!a(u_obj) !a(u_basic) about !a(u_obj)	**Boris Johnson lied about Skripal**
		John spreads rumors about Paul

The left column shows simple rules along with rule numbers, and the right column shows examples of applying these rules. The boldface part of the text examples is what would be found by the rule in the same row. Only the last phrase type, "untruth", is presented to the user. The others provide partial solutions. Note that rules 8–10 have been applied to two texts each, exemplifying the expressive power of the rule language

3.8 New and Improved Rules

An advantage of the PhraseBrowser rule language as well as other similar rule languages, is that they allow a user to quickly add rules that capture new themes of interest. The ability to modify the rules is an important feature of the tool in situations when the predefined themes do not match a specific topic and there are time constraints involved. Quickly adding a first version of a new theme is as easy as adding a word list containing some keywords. Although the obvious aim is to capture precisely everything that is relevant to a specific investigation, it is better to be able to retrieve at least some amount of desirable information through the inclusion of new rules, than risk ending up with insufficient data or no data at all. At the same time it is important to minimize the retrieval of noise, e.g., irrelevant data, which risks to burden the analysts tasked with interpreting the results.

In the present observational study the analysts worked with a set of predefined phrase types, and could not alter them or add new ones. If they had had that possibility, after studying the data they would perhaps have added phrase types for "poisons" and "lab results" to capture more of what was written about those topics. They could then, for example, easily have gone on to study what was written about persons or untruths mentioned in connection to those topics using the predefined phrase types.

While studying the results of using a first version of a rule, it is quite common to realize ways to adapt and extend it. Interesting phrases that appear in the texts that were captured can be added to the rule. Rules that capture irrelevant phrases can be altered using different features of the rule language. To this end, a user interface for interactive development of rules has been implemented for PhraseBrowser. This interface allows the analyst to iteratively refine the phrase type to capture more relevant data.

If a specific information requirement occurs in several investigations or over time, more effort could be put into studying the resulting data to improve the information gathering. Any number of methods could be used, including creating more sophisticated rules or training a machine learning model suited for the task. The goal is always to assist the analysts, and the rule language helps to put a first attempt in place quickly. A more sophisticated method (based on rules, machine learning, or any other method) can always be combined with the other phrase types to allow for varied ways to study the data. Also, whenever a new phrase type is added, many new combinations become possible. These combinations as well as the separate phrase types can be invaluable when a new subject needs to be studied.

3.9   PhraseBrowser Interface

Figure 3 shows the PhraseBrowser interface. Here 1,140,608 tweets have been downloaded using the Twitter search query "Skripal,skripal", as can be seen in the top gray area. In the left part (the blue area entitled "Phrases") the phrase type "Person" has been chosen and in the list of identified persons "Boris Johnson" is selected (the line is shaded). The tweets in the right part consequently all contain "Boris Johnson". Should any of the tweets seem interesting, it is possible to read them in context on Twitter by clicking on them.

Both the list of phrases and the list of tweets are usually much longer than what can be seen in the interface without scrolling through the lists. The count displayed next to the phrases denotes the number of tweets the phrase appears in, and the number presented before each tweet denotes the number of retweets.

To the left of each phrase in the left part of the user interface there are two buttons. Using the right button the user can plot the number of appearances of the phrases over time (not shown in this book chapter). When pressing the left button for a phrase a filter temporarily removing all tweets not containing that particular phrase is applied, which also removes all phrases not co-occurring with that particular phrase. In the example in Fig. 3 all phrases of type "Explicit Untruth" have been applied as filter, meaning that all displayed tweets contain both "Boris Johnson" and a phrase of the type "Explicit Untruth". This filter is shown in the top gray area as "All Untruth".

Several filters can be used at the same time. For instance one could go on and filter on "Boris Johnson" in addition to "All Untruth" and then look at the phrase type "Location" to see which locations are mentioned in tweets containing both of

All Untruth ✖

From: 2017-01-01 00:00:00   To: 2020-01-01 00:00:00   ⊙ Go

Phrases   ⟳   Tweets   Graph   Saved Tweets

Person   ▾   Tweet filter

☐ Auto refresh

Phrase filter		1	Text ↗	Date
Phrase	**Count**		Theresa May&Boris Johnson are so upset that there version of Skripal narrative "highly likely"a grief counsellor may be needed or EVIDENCE!	2018-04-20 21:37:32
▼ ⊿ All	25859	14	RT @veteranstoday: Boris Johnson and the Big Lies aimed at Russia - https://t.co/FlReV5g9x7 - Yesterday I did an interview with Don Grahn a…	2018-04-19 20:49:57
▼ ⊿ Boris Johnson	4050			
▼ ⊿ Ian56789	2918	3	RT @ukgranddad: @mellysbelly007 @JulieHa28304382 May's lies are dangerous. (I include Boris Johnson's lies as he appointed him Foreign Sec…	2018-04-19 17:44:38
▼ ⊿ Theresa May	1457			
▼ ⊿ Boris	677	1	@mellysbelly007 @JulieHa28304382 May's lies are dangerous. (I include Boris Johnson's lies as she appointed him For… https://t.co/Zqq5zqOajj	2018-04-19 15:50:49
▼ ⊿ theresa_may	528			
▼ ⊿ RT UK	523	43	RT @Ian56789: Boris Johnson is LYING his ass off AGAIN. #Novichok can be made almost anywhere. The #Skripal poisoning was some kind of #Fal…	2018-04-16 23:09:01
▼ ⊿ Sergei	507			
▼ ⊿ Tony Blair	427	1	RT @theforeverman: Did Boris Johnson lie that lab told him Russia was source of Salisbury nerve agent? (WATCH VIDEO) — RT. (Is the pope cat…	2018-04-15 22:09:54
▼ ⊿ Johnson	403			
▼ ⊿ Sergei Skripal	385			

Fig. 3 The PhraseBrowser user interface. The gray area at the top is the filter area, showing the Twitter search query, the number of downloaded tweets, as well as any filters that are active. In the blue "Phrases" area to the left the user can choose a phrase type ("Person" in the figure), and the corresponding phrases are displayed at the bottom. The "Tweets" area to the right displays the tweets that contain the chosen phrases in the blue area (filtered by the filters in the gray area). Both the "Phrases" and the "Tweets" areas have filter search boxes that allow the user to search among the phrases and tweets

these filters. Each filter can also be switched to its opposite, so that, for instance, only tweets not containing "Boris Johnson" are shown.

The example in Fig. 3, as described in the previous paragraphs, shows one snapshot of the work an analyst could use the tool for. By filtering on "All Untruth" he/she solely gets tweets containing explicit untruths. These tweets may be interesting since they are likely to contain accusations regarding the truthfulness of statements centered around the Twitter search topic. To get more perspectives on the accusations, the analyst could use several of the other phrase types. For instance he/she would likely use the phrase type "General Phrases" to obtain an overview of the content. In Fig. 3 the phrase type "Person" has been used, so that the list of phrases to the left shows person names that have been seen in tweets containing the accusations. For some reason the analyst takes particular interest in "Boris Johnson" here, and reads the tweets containing both this name and explicit untruths. This may provide an insight that leads the analyst to follow up by other means or use other phrase types to browse the data.

4 Method

This section describes research design, observational study setup and execution, as well as contextual factors that were present at the time of the study.

4.1 Research Design

The observational study participants ($N = 8$) were evenly split into two teams. One team was equipped with their usual set of tools, and the other team also had their ordinary tools, but was in addition provided with the PhraseBrowser text analysis tool. Hence, the general methodological approach was to collect data and compare the results of the two teams. The prototype tool had previously been used and evaluated at the unit by their personnel for some time.

The goal was to find a group of study participants that could be assumed to benefit a great deal from the use of text mining tools in their ordinary line of work. To this respect a military unit from the Swedish Armed Forces agreed to participate in the study. One of the normal tasks of the unit is to make intelligence assessments of the information environment. The participants consisted of active duty and reserve officers, as well as permanent and part-time soldiers and civilians. The aim was to put together uniform teams with respect to their educational level, experience, and gender. The team compositions were suggested by a seasoned analyst from the military unit, who consequently had some knowledge about the research design, and therefore was not allowed to personally participate actively in the observational study. The teams were constituted as indicated in Tables 4 and 5.

Table 4 Team without the PhraseBrowser tool

Age	Gender	Military category	Work role/title	Months on job	Educational level
34	Male	Civilian	Section chief	30	B.Sc.
35	Male	Civilian	Analyst	24	B.Sc.
34	Female	Civilian	Analyst	12	M.Sc.
27	Female	Civilian	Analyst	15	M.Sc.

The designated team leader in the observational study is enlisted as the topmost person in the table

Table 5 Team with the PhraseBrowser tool

Age	Gender	Military category	Work role/title	Months on job	Educational level
26	Female	Civilian	Analyst	18	B.Sc.
29	Male	Civilian	Analyst	3	M.Sc.
37	Female	Reserve	Analyst	24	University
23	Male	Soldier	Analyst	48	High school

The designated team leader in the observational study is enlisted as the topmost person in the table

4.2   Intelligence Assessment Task

The analytical questions for the teams concerned psychological operations. The case of the alleged poisoning of the former Russian intelligence officer Sergei Skripal and his daughter Yulia was used to provide a realistic backdrop for the intelligence assessment questions. This event occurred on March 4, 2018 in Salisbury, England.

As the observational study task evolved around a Psyops scenario it was assumed that text analysis would be helpful mainly for *source* and *content* analysis (see Sect. 2.2). For *source* analysis it was hypothesized that PhraseBrowser may be of use in the search for actors and authors of information used in a Psyops campaign. For *content* analysis it was hypothesized that PhraseBrowser can help extract factual information related to a campaign. The following questions/tasks were phrased:

1. Identify and document alternative explanations that contradict the British explanation.
2. Identify possible explanations and messages that might be part of a Russian influence campaign.
3. Identify and document possible sources and routes for dissemination of pro-Russian messages according to question/task 2 above.

The two teams were asked to produce an intelligence assessment consisting of a maximum of three sheets of A4 paper. They were instructed to collect data from Swedish or English language information resources only. Every information element used in the final assessment, was to be adequately referenced and, if possible, easily retrievable.

4.3   Study Setup

The observational study was conducted in situ at the military unit. The two teams resided in rooms that are normally used for similar (intelligence) work. The observational study was carried out during a single day, with additional data collection being done the day after.

The Twitter data used was selected according to the principle of relevance sampling [34]. The search keywords "Skripal,skripal" were used, which were judged to be discriminatory enough to capture all relevant tweets concerning the poisoning event. The search query resulted in a data set of 1,140,608 tweets that was downloaded through the Twitter Streaming API between March 19 and April 23, 2018, meaning that the data collection began some days after the event.

To answer the research questions, two main types of data collection were carried out: (1) the subjective views of the usefulness of the software [16] were collected from the personnel who participated in the observational study, and (2) the objective views of their performance were judged by external observers, i.e., the rating of the quality of the intelligence assessment deliverables. Observations of the work conducted by the teams were also made. Hence, the data was collected in four different ways:

1. The team members answered questions about the perceived usefulness of PhraseBrowser.
2. Four (4) subject matter experts, SMEs, were asked to judge the overall quality of the deliverables. They were to highlight what they saw as interesting pieces of information. The experts came from two different departments within the Swedish Armed Forces Headquarters and an independent think tank. Their experience from working as Russia analysts were 8, 10, 10, and 23 years.
3. All the tweets that were used by the teams were put in random order, and all participants were asked to judge the value of each individual tweet. They judged each tweet with regard to its perceived value (1–4), and its perceived area of use, e.g., (1) whether it added new hitherto unknown information, and/or, (2) if it served as a pointer to other related information sources, and/or, (3) if it (the account) was judged to be a new source itself.
4. One senior researcher per team was deployed to observe the work processes.

4.4 Observational Study Execution

The teams received the intelligence task simultaneously, and at the same time they were also given time for preparation. During this 1 h preparation, they had to designate a team leader, make an overall working plan, and make a time schedule. At least one person per team was required to read through the background material on the Skripal case from Wikipedia.[6]

Both teams were allowed to use all available information sources. Both looked into other sources than Twitter. Both teams used web browsers checking sites like google.com, hashtags.org, tweetdeck.twitter.com, etc. They also read the official government sites of various countries and traditional news outlets.

5 Results

In this section the data that was collected and the observations that were made during and after the observational study are presented.

---

[6]https://en.wikipedia.org/wiki/Sergei_Skripal [May 2, 2018].

5.1 Perceived Usefulness

The analysts who used PhraseBrowser expressed that they liked to use the tool in general. Among the perceived benefits they noted that it can be useful as events unfold in an area of interest. The tool was judged to speed up the whole data collection phase of an intelligence task. Another recurring theme in the observations about its usefulness was the breadth of the collection that enables the analyst to discover a multitude of perspectives, and allows the analyst to get an overview of an issue quickly. One respondent specifically pointed out that one of the strengths of the tool was that it steers analysts to look at data without the restraints of their preconceptions. Another useful feature of the text analysis tool that was voiced, was that it added valuable perspectives from apparently private citizens (and their accounts) in addition to the more readily available and accessible information from mainstream news and information outlets.

5.2 Subject Matter Expert Evaluation

The four SMEs, who normally read material in other formats that are not put together under time constraints, had some general remarks about the format of the deliverables, namely that they were short and not very well laid out. The SMEs also pointed out that they knew most of the information already, even though there were pieces of information that they had no prior knowledge about. Three of the four experts stated that the deliverable produced by the PhraseBrowser-team consisted of more alternate explanations and was "more detailed". Otherwise, the experts did not find any significant differences between the deliverables.

5.3 Information Fragment Value

The two teams drew information from a total of 25 unique tweets that contributed to their assessments. The team with PhraseBrowser used 15 tweets, and the team without the tool found ten relevant tweets that they used. The sample size of the data that was collected was too small to draw any general conclusions. However, some inferences can be made based on the collected data:

1. The ratio of very valuable tweets (rated with a score of 4), was almost equal between the team with PhraseBrowser (50%) and the team without it (47%).
2. At least one member of the team without PhraseBrowser stated that 90% of the tweets that they used brought hitherto unknown information. The corresponding number for the team with PhraseBrowser was 67%.
3. The members of the team with PhraseBrowser reported that the tweets that they used pointed to other potential sources to a larger extent than the other team.

5.4 General Observations of Work

It could be observed that the participants were well trained and had a common understanding of the work process, and worked according to a well-established staff methodology, which meant that they could quickly start working together and act as a team. Concerning working procedures, many similarities between the two groups could be seen with regard to how they approached the task and planned their work.

Both groups planned the work according to roughly the same schedule with two major work shifts with an interim discussion in between, and then a final synthesis discussion before the preparation of the final intelligence report. At the planned interim synthesis discussions, the analysts told what they had found as a basis for being able to draw common conclusions and align the further work. After the final work shift, longer discussions were held focusing on the possible explanations for the poisoning of Skripal, as a basis for writing the end intelligence report.

A whiteboard was used to note the conditions for the task with regard to available time and preconditions regarding limitations, intelligence/information, success factors, and immediate actions to be taken. For sharing joint work and working together, a large screen that everyone could easily see was used. Before lunch, plans were made regarding who should monitor which sources, the cutoff time for further information collection, and how the work was going to be logged.

During the afternoon the team leaders led the work and moderated the discussions that needed to be conducted. Discussions included, for example, which sources and objects that ought to be prioritized and monitored, and information sharing among the analysts concerning, for example, different spellings and synonyms to be used for the information collection ("Skripal", "Julia", and "Sergei" are equally interesting words to look for). During the work alternative explanations of the cause of the poisoning of Skripal were listed on the whiteboard along with preliminary intelligence confidence levels. These confidence levels were successively updated during the exercise based on which and how many sources that spoke for or against the respective explanations.

Both teams were judged to work systematically and be led by competent team leaders. They followed their work plans very well. Due to the severe time constraints there was some amount of stress in both teams, but the atmosphere was calm and professional. Both teams planned for and had a short 15-min break during the observational study, meaning that they invested an equal amount of time in solving the task.

5.5 Observations of Work Related to the Use of PhraseBrowser

The team with PhraseBrowser chose to designate one of the most proficient users of PhraseBrowser as team leader, which resulted in that two individuals with limited experience to operate the tool were tasked to do so. It was noted that they did not take advantage of all of the useable features of the software, and that they could not operate the software in an optimal manner. Otherwise, no significant differences in the working conditions between the teams were observed. Members of both teams explicitly strived to only take information that was collected during the observational study into account, effectively suppressing their prior knowledge of the case.

The team with PhraseBrowser used the tool to find different explanations of the Skripal incident that were proposed in the Twitter data, and to some extent to also understand whether these explanations were used in some information operation. The latter is obviously very difficult, and proved too hard to achieve during the short time allotted. The team used the tool similarly to the description in Sect. 3.9, trying several different phrase types and the filter functionality. Some of the phrase types proved useful, such as "General Phrases", "Entities", and "Explicit Untruths", as described in Sect. 3.3. Among other phrase types that were experienced as useful was one that tries to capture citations and statements, one trying to find accusations, and a few more specific phrase types for finding expressions of aggressions and tensions between different actors.

For each phrase type the team looked at phrases in order of how many times they had been used. When finding an interesting phrase, they sometimes looked at a graph of how often it had been used over time. For each interesting phrase the team always read one or more tweets these phrases appeared in. They collected the tweets that were interesting enough, taking the number of retweets into account. If these tweets had a link in them they followed that link, sometimes resulting in useful longer media articles.

6 Discussion

In this section the theory for measuring perceived usefulness as well as some validity aspects of the study are discussed. A few notes about OSINT as a data source for text mining are also provided, and the section concludes with some proposals and ideas for improvement of the PhraseBrowser tool.

6.1 Theory

This case study has aimed to examine *usefulness*, a subset of the overarching problem of user acceptance—why people embrace or reject computers and computer software. It has been shown that there are numerous variables that affect this acceptance. The widely cited technology acceptance model, TAM, conceived by Davis et al. [17], is used to model user acceptance. Davis [16] divides the variables that affect acceptance into two main categories, the *perceived ease of use* and the *perceived usefulness*. The perceived ease of use is defined by Davis [16, p. 320] as the "degree to which a person believes that using a particular system would be free of effort", or in other words: how easy it is to use a particular system. By contrast, perceived usefulness is defined as "the degree to which a person believes that using a particular system would enhance his or her job performance" [16, p. 320]. Later Venkatesh and Davis [53] presented an extended model, that they called TAM2, that among other developments, e.g., the introduction of social influence processes, divided the concept of perceived usefulness into four factors: job relevance (to what extent the proposed system is able to support one's job), output quality (how well the system performs), result demonstrability (to what extent performance can be attributed to the system), and the perceived ease of use, which in TAM2 is a direct determinant for perceived usefulness as well as for the intention to use, and ultimately—user behavior. With the extended model the performance dimension was emphasized. In this study the aim has primarily been to investigate the job relevance and the output quality aspects in terms of this model. The result demonstrability aspect was not emphasized, and ease of use questions were not considered at all.

6.2 Validity of the Study

The usefulness, or functional validity of some solution, is hard to measure because it is highly context dependent. A viable option is to compare the solution with another competing solution [34], which is what was intended here. A case study is by design a "small-N" study [21] that cannot be expected to provide conclusions with statistically significant results. On the other hand, case studies have other strengths, such as that they can provide valuable insights that may be used as a basis for further research.

The participants were operational personnel who ranked the observational study task to be on average realistic (scale: completely unrealistic/somewhat realistic/realistic/very realistic) compared to their normal tasks, and its relevance to be high to very high (scale: limited/some/high/very high). In this respect, i.e., operational personnel solving realistic tasks in a realistic setting, the results ought to be judged to have a high ecological validity [11].

A threat to the validity of the results, however, is the prior knowledge of the case by the observational study participants which could have affected the results. It was inevitable that the teams used their background knowledge of the case, even though they actively tried to avoid doing so. Therefore an effort was made to establish the level of participant background knowledge of the Skripal case. Some familiarity of the case was expected, but the main intention was to make sure that the two teams overall had a reasonably equal level of experience. Five of the eight participants stated that they had followed the events of the case briefly, while three stated that they had followed them closely (scale: not at all/briefly/closely/very thoroughly). Three of the participants answered that they had followed the case in other languages than English (e.g., in Russian and German).

There are numerous factors that affect team performance. To reach reliable conclusions in the observational study it was important to account for such contextual factors, and isolate the variables of interest, i.e., to the extent possible the variables related to the contribution of the text analysis tool only. Among several models that seek to analyze performance shaping factors to minimize human errors and optimize performance [3], the cognitive reliability and error analysis method (CREAM) of Hollnagel [28] lists such factors. The CREAM model lists "common performance conditions", in three categories: human, technological and organizational. Here we extract the identified factors of importance from the CREAM framework. CREAM suggests that eleven factors are of importance: (1) availability of resources, (2) training and experience, (3) quality of communication, (4) human-machine interface and operational support, (5) access to procedures and methods, (6) conditions of work, (7) number of goals and conflict resolution, (8) available time (time pressure), (9) circadian rhythm, (10) crew collaboration quality, and (11) quality and support of organization.

Equalization of the CREAM factors for the teams to the widest possible extent was made. The teams were given the same conditions except for factors 1 (available resources), and 4 (human-machine interface and operational support), where one team was given PhraseBrowser, and the other one was not. Here it can be noted that it was both unexpected and unfortunate that the team equipped with PhraseBrowser did not chose to assign the most proficient PhraseBrowser user(s) to operate the software. If the software would have been used more to the full extent of its capabilities, perhaps a bigger difference between the teams would have been the result. However, the research team decided not to interfere with the internal work processes, i.e., the division of labor, of either team. It was not possible to make factor 2 (training and experience) identical for the two teams, but a justification for the team composition has been presented in this chapter. Factor 8 (available time) was the same for both teams, but the participants stated that (on average) at least four times as much time would be needed to solve the given task sufficiently well. Thus, it is reasonable to assume that the limited time affected the quality of the outputs for both teams negatively but equally. Another aspect with regard to time that may have affected the results, is that the data collection of tweets started only some time, i.e., around 2 weeks, after the incident took place.

6.3   Open Source Intelligence as a Data Source for Text Mining

Since the main information source for the intelligence assessments was OSINT, a few words on OSINT as a source is in place. One of the drawbacks that was found some time ago and that has already been mentioned, is that the sheer amount of data can be overwhelming for individuals or even organizations [29]. Moreover, even if a huge amount of data is easily accessible, much data still remain out of reach for practitioners due to various constraints, e.g., lack of access to closed forums, pay-for services, and password-protected sites that are prohibited, and there are also other legal constraints such as copy restrictions that may hinder access [29]. Another drawback of publicly available information is that the quality of it can be questionable—after all, anyone can post just about anything on the Internet without discrimination.

A significant part of the OSINT problem is how to handle unstructured, soft data, and Dragos [19] noted that there are multiple other uncertainties beside the credibility of sources in such data, for example, intrinsic properties such as the ambiguity of natural language and the presence of possible inconsistencies in the provided information. On a smaller scale, practical obstacles such as the handling of several obscure languages and the prevalence of "slang" language pose problems in the information organization work phase when it comes to unifying and cleaning the data for further processing [29].

6.4 Potential Improvements of PhraseBrowser

During and after the observational study, the participating analysts provided several suggestions on how PhraseBrowser could be improved. Some of these thoughts are related to and discussed here, along with a discussion on the improvements that have been made since the observational study, as well as some plans for future work.

The analysts found the phrase types and phrases useful. However, as there are already many phrase types to choose from, it is not always easy to understand what all of them were created to support. The analysts would have liked to have short explanations of the phrase types, and also some indication about how well-developed they are. The latter could partially be answered by displaying the number of rule lines in the phrase types.

The participants also realized the potential benefits of being able to create their own phrase types, and asked for this functionality. Since then a first version of a user interface for rule development has been implemented. To be expressive, the rule language is still, however, somewhat complex, so only dedicated analysts, perhaps with some computer science knowledge, can be expected to use it to its full potential.

PhraseBrowser is focused on providing an overview of the textual content in tweets. The analysts would have liked to have access to more sources within the interface, such as other social media platforms, news media, etc. Since then partial support for RSS feeds has been implemented.

During the study the analysts found interesting tweets that led them to interesting Twitter accounts. They would have liked to get more information about these accounts within the prototype, and be able to search for specific accounts in the data. This functionality will possibly be implemented in a separate, complementary prototype.

There are many ways to filter data that are implemented in the PhraseBrowser prototype, some of which were also mentioned by the analysts, such as URLs and location information derived from accounts. It is also possible to look at data during smaller or larger time intervals and for chosen phrases, but the analysts would have liked this functionality to be more developed.

As described in Sect. 3.4, PhraseBrowser does not necessarily analyze every tweet when the data stream is too large compared to the computing power. Also, using phrases as filters, as described in Sect. 3.3, reduces the set of tweets under investigation. These facts are currently reflected to some extent in the user interface, but the analysts would have liked them to be more prominent.

PhraseBrowser, and the set of prototype tools it is part of, allows for a great deal of flexibility. After using the possibilities for a while, certain usage patterns might emerge. For instance, it could under certain circumstances prove beneficial to filter data by "Explicit Untruths" and a particular list of persons, before trying to explore what is written. If further methods for filtering data, such as by image content or user account meta data, are available, even more complex usage patterns may prove useful. Such usage patterns could potentially be executed in advance, with the results presented using a digital dashboard.

7 Conclusions

This chapter presents the results of a case study that seeks to investigate the perceived usefulness of a text mining tool and how it affects the quality of the end result (output) of a realistic intelligence assessment task. Conclusions from a case study like this, however, should be interpreted in light of the limited scope of the study, e.g., the number of participants and the study design at large, and be regarded as preliminary. As outlined in Sect. 1, two research questions have governed the present study:

– Is PhraseBrowser, a specific instantiation of a text analysis tool, perceived as useful for solving typical analytical tasks?

It was found that all analysts that used or had used the PhraseBrowser tool previously, *liked* the tool and subjectively *perceived* it as useful. The main benefit was that it was thought to provide analysts with an opportunity to get an overview of an issue quickly. The results indicate that its main contribution is to highlight pointers to other sources making it possible to conduct further searches, and to a lesser extent also to find unique information and new Twitter sources.

– Does the use of the text analysis tool improve the quality of a typical analytical deliverable?

Based on the inputs from the study participants and the SMEs, it was not possible to discern any major quality differences in the intelligence assessment deliverables. The assessment of the deliverable from the team that operated the tool reported that it contained more diverse and "more detailed" pieces of information than the deliverable from the team that did not have the tool.

In the future the research methodology ought to be developed further, and more observational studies with other groups of analysts should be undertaken. A specific observational study could be to examine the usefulness of the PhraseBrowser tool for intelligence analysis of ongoing events. It should also be noted that the value of PhraseBrowser was examined relative to the use of a range of other pieces of software, e.g., the ones that the team that did not use the tool had at its disposal. To further strengthen the results of this study, future research should strive to also establish the function of the additional software used, in some detail.

Acknowledgements The authors would like to express their gratitude to the commander of the military unit that took part in the observational study, and especially to the enthusiastic and highly professional personnel involved in the preparation and execution of the study. This work was supported by the Swedish Armed Forces, and by the European Union Horizon 2020 program (grant agreement no. 832921).

References

1. W. Agrell, G.F. Treverton, The science of intelligence: reflections on a field that never was, in *National Intelligence Systems: Current Research and Future Prospects*, ed. by G.F. Treverton, W. Agrell, Chap. 11 (Cambridge University Press, New York, 2009), pp. 265–280
2. B. Alex, C. Grover, B. Haddow, M. Kabadjov, E. Klein, M. Matthews, S. Roebuck, R. Tobin, X. Wang, Assisted curation: does text mining really help?, in *Proceedings of the Pacific Symposium on Biocomputing (PSB 2008)* (World Scientific, Singapore, 2008), pp. 556–567. https://doi.org/10.1142/9789812776136_0054
3. M.A.B. Alvarenga, P.F.F. Frutuoso e Melo, R.A. da Fonseca, A review of the models for evaluating organizational factors in human reliability analysis, in *Proceedings of the 2009 International Nuclear Atlantic Conference (INAC 2009)* (Brazilian Nuclear Energy Association (ABEN), Rio de Janeiro, 2009)
4. L. Armistead (ed.), *Information Operations: Warfare and the Hard Reality of Soft Power*, Issues in Twenty-First Century Warfare (Brassey's, Washington, 2004)
5. S. Arnborg, J. Brynielsson, H. Artman, K. Wallenius, Information awareness in command and control: precision, quality, utility, in *Proceedings of the Third International Conference on Information Fusion (FUSION 2000)*, vol. 2 (IEEE, Piscataway, 2000), pp. ThB1/25–32. https://doi.org/10.1109/IFIC.2000.859871
6. M. Bang, Military intelligence analysis: institutional influence. Ph.D. thesis, National Defence University, Helsinki, October 2017. http://urn.fi/URN:ISBN:978-951-25-2930-80
7. M.T. Bimfort, A definition of intelligence. Stud. Intell. **2**(4), 75–78 (1958)
8. M. Boicu, G. Tecuci, D.A. Schum, Intelligence analysis ontology for cognitive assistants, in *Proceedings of the Third International Ontology for the Intelligence Community Conference (OIC-2008)* (George Mason University, Fairfax, 2008), pp. 31–35
9. K. Bontcheva, L. Derczynski, A. Funk, M.A. Greenwood, D. Maynard, N. Aswani, TwitIE: an open-source information extraction pipeline for microblog text, in *Proceedings of the Ninth International Conference on Recent Advances in Natural Language Processing (RANLP 2013)* (Association for Computational Linguistics, Stroudsburg, 2013), pp. 83–90
10. A. Breakspear, A new definition of intelligence. Intell. Natl. Secur. **28**(5), 678–693 (2013). https://doi.org/10.1080/02684527.2012.699285
11. U. Bronfenbrenner, Toward an experimental ecology of human development. Am. Psychol. **32**(7), 513–531 (1977). https://doi.org/10.1037/0003-066X.32.7.513

12. J. Brynielsson, A. Horndahl, L. Kaati, C. Mårtenson, P. Svenson, Development of computerized support tools for intelligence work, in *Proceedings of the 14th International Command and Control Research and Technology Symposium (14th ICCRTS)* (U.S. Department of Defense CCRP, Washington, 2009). Paper no. 48

13. Cisco, Cisco visual networking index: forecast and trends, 2017–2022. White paper C11-741490-00, February 2019

14. R.M. Clark, *Intelligence Analysis: A Target-Centric Approach*, 5th edn. (CQ Press, Washington, 2016)

15. H. Cunningham, V. Tablan, A. Roberts, K. Bontcheva, Getting more out of biomedical documents with GATE's full lifecycle open source text analytics. PLOS Comput. Biol. **9**(2), e1002854 (2013). https://doi.org/10.1371/journal.pcbi.1002854

16. F.D. Davis, Perceived usefulness, perceived ease of use, and user acceptance of information technology. MIS Q. **13**(3), 319–340 (1989). https://doi.org/10.2307/249008

17. F.D. Davis, R.P. Bagozzi, P.R. Warshaw, User acceptance of computer technology: a comparison of two theoretical models. Manag. Sci. **35**(8), 982–1003 (1989). https://doi.org/10.1287/mnsc.35.8.982

18. M. Degaut, Spies and policymakers: intelligence in the information age. Intell. Natl. Secur. **31**(4), 509–531 (2016). https://doi.org/10.1080/02684527.2015.1017931

19. V. Dragos, An ontological analysis of uncertainty in soft data, in *Proceedings of the 16th International Conference on Information Fusion (FUSION 2013)* (IEEE, Piscataway, 2013), pp. 1566–1573

20. M.J. Eppler, J. Mengis, The concept of information overload: a review of literature from organization science, accounting, marketing, MIS, and related disciplines. Inf. Soc. **20**(5), 325–344 (2004). https://doi.org/10.1080/01972240490507974

21. A.L. George, A. Bennett, *Case Studies and Theory Development in the Social Sciences* (MIT Press, Cambridge, 2005)

22. M. Glassman, M.J. Kang, Intelligence in the internet age: the emergence and evolution of open source intelligence (OSINT). Comput. Hum. Behav. **28**(2), 673–682 (2012). https://doi.org/10.1016/j.chb.2011.11.014

23. J.K. Guo, D. Van Brackle, N. LoFaso, M.O. Hofmann, Extracting meaningful entities from human-generated tactical reports. Procedia Comput. Sci. **61**, 72–79 (2015). https://doi.org/10.1016/j.procs.2015.09.153

24. W.M. Hall, G. Citrenbaum, *Intelligence Analysis: How to Think in Complex Environments* (Praeger, Santa Barbara, 2009)

25. W. He, S. Zha, L. Li, Social media competitive analysis and text mining: a case study in the pizza industry. Int. J. Inf. Manag. **33**(3), 464–472 (2013). https://doi.org/10.1016/j.ijinfomgt.2013.01.001

26. M.A. Hearst, Automatic acquisition of hyponyms from large text corpora, in *Proceedings of the 15th International Conference on Computational Linguistics (COLING-92)*, vol. 2 (Association for Computational Linguistics, Stroudsburg, 1992), pp. 539–545

27. R.J. Heuer, Jr., *Psychology of Intelligence Analysis* (Center for the Study of Intelligence, Central Intelligence Agency, Washington, 1999)

28. E. Hollnagel, *Cognitive Reliability and Error Analysis Method (CREAM)* (Elsevier, Oxford, 1998). https://doi.org/10.1016/B978-0-08-042848-2.X5000-3

29. A.S. Hulnick, The downside of open source intelligence. Int. J. Intell. CounterIntell. **15**(4), 565–579 (2002). https://doi.org/10.1080/08850600290101767

30. J. Jiang, Information extraction from text, in *Mining Text Data*, ed. by C.C. Aggarwal, C. Zhai, Chap. 2 (Springer, Boston, 2012), pp. 11–41. https://doi.org/10.1007/978-1-4614-3223-4_2

31. F. Johansson, J. Brynielsson, P. Hörling, M. Malm, C. Mårtenson, S. Truvé, M. Rosell, Detecting emergent conflicts through web mining and visualization, in *Proceedings of the 2011 European Intelligence and Security Informatics Conference (EISIC 2011)* (IEEE, Piscataway, 2011), pp. 346–353. https://doi.org/10.1109/EISIC.2011.21

32. D. Jurafsky, J.H. Martin, *Speech and Language Processing: An Introduction to Natural Language Processing, Computational Linguistics, and Speech Recognition*, 2nd edn. (Prentice Hall, Upper Saddle River, 2009)
33. C. Koopman, J. Snyder, R. Jervis, Theory-driven versus data-driven assessment in a crisis: a survey of international security readers. J. Confl. Resolut. **34**(4), 694–722 (1990). https://doi.org/10.1177/0022002790034004006
34. K. Krippendorff, *Content Analysis: An Introduction to Its Methodology*, 2nd edn. (SAGE, Thousand Oaks, 2004)
35. T.K. Landauer, *The Trouble with Computers: Usefulness, Usability, and Productivity*, A Bradford Book (MIT Press, Cambridge, 1995)
36. E.V. Larson, R.E. Darilek, D. Gibran, B. Nichiporuk, A. Richardson, L.H. Schwartz, C.Q. Thurston, *Foundations of Effective Influence Operations: A Framework for Enhancing Army Capabilities* (RAND Corporation, Santa Monica, 2009). https://www.rand.org/pubs/monographs/MG654.html
37. K. Lim, Big data and strategic intelligence. Intell. Natl. Secur. **31**(4), 619–635 (2016). https://doi.org/10.1080/02684527.2015.1062321
38. D. McDonald, U. Kelly, The value and benefit of text mining to UK further and higher education. Digital Infrastructure Directions Report, Doc# 811, Version 1.1, JISC, March 2012. https://www.jisc.ac.uk/reports/value-and-benefits-of-text-mining
39. NATO, Allied joint doctrine for psychological operations. Publication AJP-3.10.1(A), NATO Standardization Agency, Brussels, October 2007
40. D. Omand, J. Bartlett, C. Miller, Introducing social media intelligence (SOCMINT). Intell. Natl. Secur. **27**(6), 801–823 (2012). https://doi.org/10.1080/02684527.2012.716965
41. J.K. Pal, Usefulness and applications of data mining in extracting information from different perspectives. Ann. Libr. Inf. Stud. **58**(1), 7–16 (2011)
42. N.A. Pollard, On counterterrorism and intelligence, in *National Intelligence Systems: Current Research and Future Prospects*, ed. by G.F. Treverton, W. Agrell, Chap. 6 (Cambridge University Press, New York, 2009), pp. 117–146
43. A.H. Razavi, D. Inkpen, R. Falcon, R. Abielmona, Textual risk mining for maritime situational awareness, in *Proceedings of the 2014 IEEE International Inter-Disciplinary Conference on Cognitive Methods in Situation Awareness and Decision Support (CogSIMA)* (IEEE, Piscataway, 2014), pp. 167–173. https://doi.org/10.1109/CogSIMA.2014.6816558
44. F.E. Ritter, G.D. Baxter, E.F. Churchill, *Foundations for Designing User-Centered Systems: What System Designers Need to Know about People* (Springer, London, 2014). https://doi.org/10.1007/978-1-4471-5134-0
45. D.E. Sichel, *The Computer Revolution: An Economic Perspective* (Brookings Institution Press, Washington, 1997)
46. R.D. Steele, *On Intelligence: Spies and Secrecy in an Open World* (AFCEA International Press, Fairfax, 2000)
47. S. Stieglitz, L. Dang-Xuan, A. Bruns, C. Neuberger, Social media analytics: an interdisciplinary approach and its implications for information systems. Bus. Inf. Syst. Eng. **6**(2), 89–96 (2014). https://doi.org/10.1007/s12599-014-0315-7
48. Swedish Armed Forces, Försvarsmaktens underrättelsereglemente [The armed forces' intelligence manual]. Publication M7739-353025, Headquarters, Stockholm, December 2010
49. S. Tzu, *The Art of War*, trans. by S.B. Griffith (Clarendon Press, Oxford, 1963)
50. U.S. Department of Defense, Joint doctrine for information operations. Joint Publication 3.13, Joint Chiefs of Staff, Washington, October 1998
51. U.S. Department of Defense, Joint and national intelligence support to military operations. Joint Publication 2-01, Joint Chiefs of Staff, Washington, July 2007
52. U.S. Department of Defense, Information operations. Joint Publication 3.13, Joint Chiefs of Staff, Washington, November 2012

53. V. Venkatesh, F.D. Davis, A theoretical extension of the technology acceptance model: four longitudinal field studies. Manag. Sci. **46**(2), 186–204 (2000). https://doi.org/10.1287/mnsc. 46.2.186.11926
54. C. von Clausewitz, *On War*, trans. by M. Howard, P. Paret (Princeton University Press, Princeton, 1976). Originally published 1832

Printed in the United States
by Baker & Taylor Publisher Services